AF789559

Adaptations of Desert Organisms

Edited by J. L. Cloudsley-Thompson

J. L. Cloudsley-Thompson

Ecophysiology of Desert Arthropods and Reptiles

With 77 Figures

Springer-Verlag
Berlin Heidelberg New York
London Paris Tokyo
Hong Kong Barcelona
Budapest

Prof. Dr. John L. Cloudsley-Thompson
University of London
University College
Department of Biology (Medawar Building)
Gower Street
London WC1E 6BT
United Kingdom

Cover illustration: photograph by J. L. Cloudsley-Thompson

ISBN-13: 978-3-642-75339-8 e-ISBN-13: 978-3-642-75337-4
DOI: 10.1007/978-3-642-75337-4

Library of Congress Cataloging-in-Publication Data. Cloudsley-Thompson, J. L. Ecophysiology of desert arthropods and reptiles / J. L. Cloudsley-Thompson. p. cm. – (Adaptations of desert organisms) Includes index. ISBN-13: 978-3-642-75339-8
1. Desert fauna – Ecology. 2. Arthropoda – Ecology. 3. Reptiles – Ecology. 4. Desert fauna – Physiology. 5. Arthropoda – Physiology. 6. Reptiles – Physiology. 7. Adaptation (Biology) I. Title. II. Series. QL 116.C58 1991 597.9'0452652 – dc20

This work is subject to copyright. All rights are reserved, whether the whole or part of the material is concerned, specifically the rights of translation, reprinting, re-use of illustrations, recitation, broadcasting, reproduction on microfilms or in other ways, and storage in data banks. Duplication of this publication or parts thereof is only permitted under the provisions of the German Copyright Law of September 9, 1965, in its current version, and a copyright fee must always be paid. Violations fall under the prosecution act of the German Copyright Law.

© Springer-Verlag Berlin Heidelberg 1991
Softcover reprint of the hardcover 1st edition 1991

The use of general descriptive names, registered names, trademarks, etc. in this publication does not imply, even in the absence of a specific statement, that such names are exempt from the relevant protective laws and regulations and therefore free for general use.

Typesetting: International Typesetters Inc., Makati, Philippines
31/3145-543210 – Printed on acid-free paper

Acknowledgments

This book embodies ideas and information emanating from discussions and correspondence with many friends and colleagues, including F.T. Abushama, G.A. Ahearn, R.A. Avery, A. d'A. Bellairs, C. Constantinou, C.S. Crawford, J.E. Dalingwater, E.B. Edney, N.F. Hadley, D. Happold, M. Happold, J. Henschel, G.N. Louw, D. Mahsberg, L. Malan, G.A. Polis, M.K. Seely, P.A. Selden, K.L.G. Smith, I.F. Spellerberg, W.G. Whitford, A.J. Wallwork and my dearest wife Anne Cloudsley.

My thanks are due to the various authors who have allowed me to reproduce figures from their publications. These are acknowledged in the figure legends. Warmest thanks are also extended to Maureen Hakney and Helen Wilson for typing the manuscript. The book was completed during tenure of a Leverhulme Emeritus Fellowship which I am pleased to acknowledge, while Professor N.A. Mitchison FRS kindly provided me with accommodation in University College London.

London, Spring 1991 J.L. CLOUDSLEY-THOMPSON

Contents

1 Introduction

The Arthropoda and Reptilia are two of the most successful taxa of desert animals. They share in common the possession of a relatively impermeable integument, excretion of insoluble nitrogenous waste with little loss of accompanying water, primarily ectothermal regulation of body temperature and consequent metabolic efficiency. They differ in having exoskeletons and endoskeletons respectively, characters that are associated with their relatively smaller and larger sizes.

The adaptations of both taxa to desert life have received detailed reviews in recent years; but the treatment accorded them in the present volume differs somewhat from that which they have previously received. Instead of exhaustively listing their physiological, behavioural and ecological specializations for desert life in general, the various parameters of the desert environment are first defined: then the ways in which these are accommodated within the two taxonomic groups are compared and their similarities emphasized. Certain topics, such as the roles of detritivores and invertebrate predators in the ecosystem, pollenivory and nectivory, metabolism, ion exchange and nitrogen excretion, have been deliberately omitted or abridged in the expectation that they will be covered more fully in forthcoming books in this Series.

Organisms that fill similar ecological niches in different, independently evolved biotas, are termed 'ecological equivalents.' Such convergent evolutionary responses to the desert environment are to be found in different continents, and can be identified by similar patterns of resource utilization. They are useful in illustrating the manner in which selection has operated in physically similar, but geographically widely separated habitats.

Although the terms 'ecophysiology', 'environmental physiology' and 'physiological ecology' are often employed almost as pseudonyms, the three disciplines are subtly distinct. Environmental physiology is primarily concerned with responses to meteorological parameters, physiological ecology is concerned with animals in their natural environments rather than in the laboratory, while ecophysiology attempts to integrate individual responses within the context of a single population or species. Were these strict criteria to be adhered to, the title of this book should really have included the words 'environmental physiology' since both this and ecophysiology have received attention.

Another point of importance is that, on account of their relatively small size and correspondingly large surface to volume ratios when compared with most birds and mammals, arthropods and reptiles of necessity have evolved low transpiration rates. They are therefore already pre-adapted for water conservation and rapid thermal exchange in the desert environment. Indeed, in some cases, the range of physiological adaptation of individual animals may even exceed in extent that between the evolutionary adaptations of related desert and mesic species. For adaptation can be both physiological and evolutionary. Physiological adaptation is the response of an individual organism as a result of exposure to certain environmental conditions, making it more resistant to those conditions. Evolutionary adaptations result from natural selection operating over a long period of time. They improve the chances of survival, in a particular environment, of the individuals possessing them. Most of the animals that inhabit the desert are to some extent adapted to the extreme conditions of their environment, at least to the extent of selecting more favourable microhabitats. The forms that inhabit less harsh biomes, on the other hand, may either be adapted to one particular type of environmental niche or they may be generalists, i.e. adapted to survive in a variety of ecological habitats.

According to Rutowski and Alcock (1989), for example, diversity rather than uniformity characterizes insect mating systems in the Sonoran desert of Arizona. Sufficient ecological variation exists to have produced a diversity of distribution patterns in females, with correlated diversity in the tactics employed by males to locate and mate with as many of these females as possible. The stark and open aspect of the desert topography is probably responsible in part for the tendency of species such as butterflies, which use landmarks to encounter males, to aggregate on mountain ridges and hilltops. These conspicuous sites, like washes, appear to act as natural orientation guides for many wide-ranging insect species. Sexual selection operates in the same fundamental pattern in both desert and non-desert environments, and there is sufficient ecological diversity in the former to have generated adaptive radiation of insect mating systems that parallels what is present in other environments throughout the world.

The common assumption underlying much of the research on reptilian ecophysiology over the past two decades, i.e. that desert reptiles are uniquely adapted to the harsh environment in which they occur, has been questioned by Bradshaw (1988) who suggested that they may be able to adjust successfully to a wide range of environmental stresses without recourse to genetic adaptation. Indeed, he concluded that few, if any, desert reptiles ever experience thermal stress in the field because of the efficacy of their thermoregulatory behaviour.

Foremost among fundamental characteristics to which Bradshaw (1988) attributed the obvious success of reptiles in desert environments around the world are the following, all of which are also found among arthropods: (1) low rates of metabolism which result in extremely low rates of resource utilization and economy in the handling of water; (2) ectothermy and behavioural thermoregulation; (3) the excretion of nitrogenous waste products in the form of relatively insoluble uric acid; (4) the ability to withstand marked perturbations of the *milieu interieur* for extended periods of time; and (5) the widespread use of avoidance behaviour during periods when continued activity would be counterproductive.

Desert ecology is not so much the ecology of organisms that inhabit arid and thermally instable environments as it is of plants and animals subjected to spatial patchiness, temporal variability and unpredictable resources (including water), competitors and predation, rather than with scarcity of water alone (Safriel et al. 1989). The challenge of the desert is not only a matter of coping with scarcity, but also of exploiting temporarily favourable conditions. It is possible, therefore, to understand why there should be so few clearly differentiated ecophysiological specializations for desert living to be found among arthropods and reptiles. Nevertheless, it would be going too far to claim that all arthropods and reptiles are equally well adapted to the desert environment; but desert animals tend to show adaptations to individual facets of that environment, depending upon their way of life, rather than to the generality of environmental factors.

Previously published major reviews of the adaptations of arthropods and reptiles to the desert environment include the following: (1) *Arthropoda*: Cloudsley-Thompson (1964a, 1970b, 1975), Crawford (1981), Edney (1974, 1977). (2) *Reptilia*: Bradshaw (1986), Cloudsley-Thompson (1971, 1988b), Mayhew (1968), Pianka (1986), Schmidt-Nielsen and Dawson (1964).

2 The Deserts of the World and Their Faunas

2.1 Classification of Deserts

The term 'desert' is imprecise: there is no general agreement as to what constitutes a desert and no precise definition of the term (Heathcote 1983). McGinnies et al. (1968) classified the world's deserts, according to mean annual precipitation, as follows:

1. Extremely arid (less than 60-100 mm);
2. Arid (from 60-100 to 150-250 mm);
3. Semi-arid (from 150-250 to 250-500 mm).

Although a classification based solely on precipitation is clearly inadequate, it does reflect the pre-eminence of water in controlling the productivity of desert ecosystems (Louw and Seely 1982). Desert rainfall is both erratic and unpredictable. Aridity is associated with considerable insolation and high temperatures during the day; while hot, strong winds, usually associated with sandstorms, are a constant feature of desert climates.

Arid environments can be created by three general factors, acting either individually or in combination. These are: (1) separation from oceanic moisture, either by distance (interior continental deserts) or mountain ranges (rain-shadow deserts); (2) the persistence of dry stable air masses that resist convective currents; (3) the lack of storm systems which would create instability and provide the lifting necessary for precipitation (McGinnies et al. 1968). These causal factors can be used to provide the following broad classification of the more important deserts of the world (Cloudsley-Thompson 1977a):

1. Subtropical and tropical deserts. These are largely the result of semi-permanent belts of high pressure and descending air masses. They include much of the Great Palaearctic, North American and Australian deserts.

2. Cool coastal deserts. Almost always rainless, these are periodically drenched with chill fogs. Rainlessness again results from descending air masses, the high humidity and cold air from nearby cool ocean currents. Examples are the Namib, Atacama and Baja California deserts which are adjacent to the Benguela, Humboldt and California currents respectively.

3. Rain-shadow deserts. Examples are the Mojave and Great Basin deserts of North America, the deserts of Patagonia and part of the Great Australian desert.

4. Interior continental deserts. Absence of marine influence and other factors related to the massive bulk of surrounding land are responsible for much of the Great Palaearctic, North American and Australian deserts.

As recognized in this book, the major deserts of the world are:

1. *The Great Palaearctic desert* (including the Sahara, Somali-Chalbi, Arabian, Iranian, Thar, Turkestan, Takla-Makan and Gobi deserts; Figs. 1,2).
2. *The North American desert* (including the Great Basin, Mojave, Sonoran and Chihuahuan deserts, Figs. 3,4).
3. *The Atacama-Peruvian desert* (Fig. 5).
4. *The Monte-Patagonian desert.*
5. *The Kalahari-Namib desert* (Fig. 6).
6. *The Australian desert* (including the Gibson, Simpson, and Great Victoria deserts; Fig. 7).

Fig. 1. a Great Palaearctic desert — Nubian desert, Sahara northern Sudan. **b** Great Palaearctic desert — central Sahara, southern Algeria

a
b

McGinnies (1979) gives a useful summary of the common features of arid-land ecosystems throughout the world.

2.2 Precipitation and Moisture

Regardless of its locality and climate, the input of moisture is the principal factor limiting the primary productivity of the flora and fauna of any desert ecosystem (Whitford 1986). From the viewpoint of the biota, therefore, the most useful classification of deserts would be one based upon the types and amounts of precipitation. Using these criteria, the world's deserts can be classified as follows:

1. Deserts with seasonal rains (including deserts with winter-Mediterranean and summer-monsoon types of rainfall).
2. Fog and mist deserts (including cool, coastal deserts where fog appears at irregular intervals throughout the year).
3. Deserts with irregular rain (including some interior continental desert regions and the Australian desert).

Whether summer rains, which ameliorate the extreme heat of the summer, or winter rains, which evaporate less rapidly, are more favourable to the fauna and flora, is far from clear. In the opinion of the author, winter rain is probably more beneficial because it does not evaporate so rapidly and has more time to percolate into the soil. The fauna is markedly influenced by unfavourable bioclimatic conditions, especially of temperature and humidity, as well as by shortage of food. Of these, high daytime temperature is probably the most important, for relative humidity is generally so low that only species capable in some way of withstanding or avoiding severely desiccating conditions in summer are able to exist at all (Cloudsley-Thompson 1962a, 1969a). The Sonoran desert as a whole is relatively verdant in comparison with areas of similar annual precipitation in the Chihuahuan desert, because the distribution of moisture is more even in the former.

Cool, coastal deserts, such as the Namib and Atacama, seldom receive appreciable rainfall. Fog and mist, however, represent an extremely important alternative source of moisture. The atmospheric humidity is high: even in the absence of fog, the air is sufficiently moist to provide for the requirements of lichens and other specialized types of vegetation. Away from the coast, however, the average relative humidity drops sharply, the temperature increases and the flora and fauna are very much reduced.

Even though rain may be irregular in most interior continental deserts, their biota is similar to, although more restricted than that of adjacent regions which

Fig. 2. a Great Palaearctic desert — Thar desert, Rajasthan. **b** Great Palaearctic desert —Wahiba sands, Oman

Fig. 3. **a** North American desert — Sonoran desert, Mexico. **b** North American desert — Sonoran desert, Arizona

Fig. 4. **a** North American desert — Chihuahuan desert, New Mexico. **b** North American desert — Great Basin desert, Nevada

Fig. 5. **a** Peruvian desert — north of Lima. **b** Peruvian desert; vegetation maintained by maritime fog

enjoy seasonal rainfall. The one desert that is quite different is the Australian desert, for here several years may pass without any precipitation at all. Consequently, *r*-selected univoltine arthropods such as Solifugae are not found, while K-selected scorpions and mygalomorph spiders flourish (Cloudsley-Thompson 1986, 1988a). Another consequence of the unpredictably heavy rainstorms of the Australian climate is occasional sheetflooding. Burrowing arthropods have countered this with a variety of almost impregnable entrances to their retreats (Sect. 6.3).

Fig. 6. **a** Kalahari desert; vegetated dune with gemsbok. **b** Namib desert; dune field bordering the dry Kuiseb river

Fig. 7 a,b. Central Australian desert — Simpson desert

2.3 Exposure

Except when they are in their retreats or burrows, desert animals are exposed not only to the exigencies of the climate, but also to the rapacity of predators. This is true of many smaller animals in other biomes, but exposure produces especially marked effects in flat desert regions where there is little vegetative cover. Desert adaptations can therefore be divided into two categories: (1) those involved at all times, both when an animal is in the open and while it is hidden; (2) those, such as adaptive colouration or the avoidance of wind, which apply only when an animal is out in the open. This point has a bearing on all the adaptations to be discussed in the following pages.

2.4 Parallel Evolution, Convergence and Ecological Equivalents

2.4.1 Reptilian Examples of Convergence

Parallel evolution, convergence and adaptive radiation are all aspects of the same phenomenon, but considered from different viewpoints. Animals that inhabit similar ecological niches, whether they be closely related taxonomically, or not, tend to acquire a similar appearance and behaviour as a result of parallel evolution. The concept of parallel evolution, convergence and ecological equivalents (with comparable morphological characters resulting from adaptation to similar environments) has already been introduced.

A dramatic example of ecological equivalents is provided by *Moloch horridus* (Fig. 8a), the thorny devil of central and western Australia, a cryptically coloured and heavily armoured agamid lizard that feeds on ants, and its ecological analogue *Phrynosoma platyrhinos* (Fig. 8b), the horned iguanid lizard of the North American deserts which likewise exploits a diet of ants. Morphological analysis demonstrates that *M. horridus* and *P. platyrhinos* are anatomically closer to one another than either species is to any other member of its own lizard taxon. It is easy, however, to over-emphasize the significance of convergence, and the biologist must be careful not to allow it to bias his judgement. Nevertheless, the concept is useful in a restricted sense. Although the approach of this book extends comparisons to two completely unrelated taxa, one object of doing so is to illustrate how similar environmental parameters have resulted in comparable adaptations.

Both North American and Australian desert lizard faunas include long-legged species that frequent open spaces between plants. Pianka (1985) cites *Callisaurus draconoides* (Iguanidae) in North America and *Ctenophorus scutulatus*, *C. cristatus* and *C. isolepis* (Agamidae) in Australia, as examples. The North American leopard lizard *Gambelia* (= *Crotaphytus*) *wislizenii* (Iguanidae) feeds on other lizards in the southern part of its geographical range: convergent with it is the Australian *Varanus eremius* (Varanidae), another medium-sized, lizard-eating species. Active, widely foraging lizards occur in all deserts of the world, e.g.

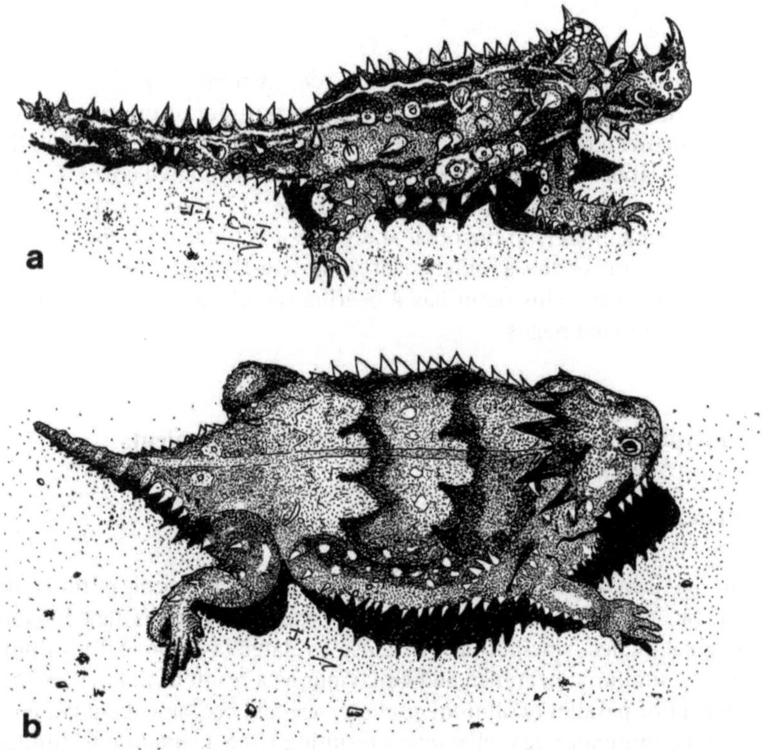

Fig. 8. **a** *Moloch horridus* (Agamidae), the thorny devil of the Australian deserts. **b** *Phrynosoma* sp. (Iguanidae), the horned toad of the North American deserts

Cnemidophorus spp. (Teiidae; Fig. 9a) in North America, *Acanthodactylus* spp. (Lacertidae; Fig. 9b) in the Great Palaearctic desert, *Ctenotus* spp. (Scincidae) in Australia and *Nucras* spp. (Lacertidae) in the Kalahari. Convergent nocturnal geckos include *Pachydactylus capensis* and the Australian *Heteronotia binoei*. The subterranean skinks *Typhlosaurus* spp. and *Lerista* spp. are likewise roughly similar in their anatomy and ecology, as are the semi-arboreal agamids, *Agama hispida* and *Pogona minor* (see also Sect. 8.4.2). Despite the existence of such similarities, Pianka (1985) concludes with a word of caution: "Ecologies of putatively convergent species pairs inevitably differ markedly when scrutinized."

Another well-known reptilian example of ecological equivalents is provided by the North American sidewinder rattlesnake (*Crotalus cerastes*; Fig. 10a) and its counterpart, the Saharan sand viper (*Cerastes cerastes*; Fig. 10b). Both species move by throwing lateral loops forward, and hide themselves by flattening their bodies and shovelling sand over their backs (Fig. 11). In general appearance they are so much alike that, were it not for the rattle of *Crotalus cerastes* and the pit between the eye and the nostril, these species might easily be confused. In contrast, there is relatively little close resemblance between the North American

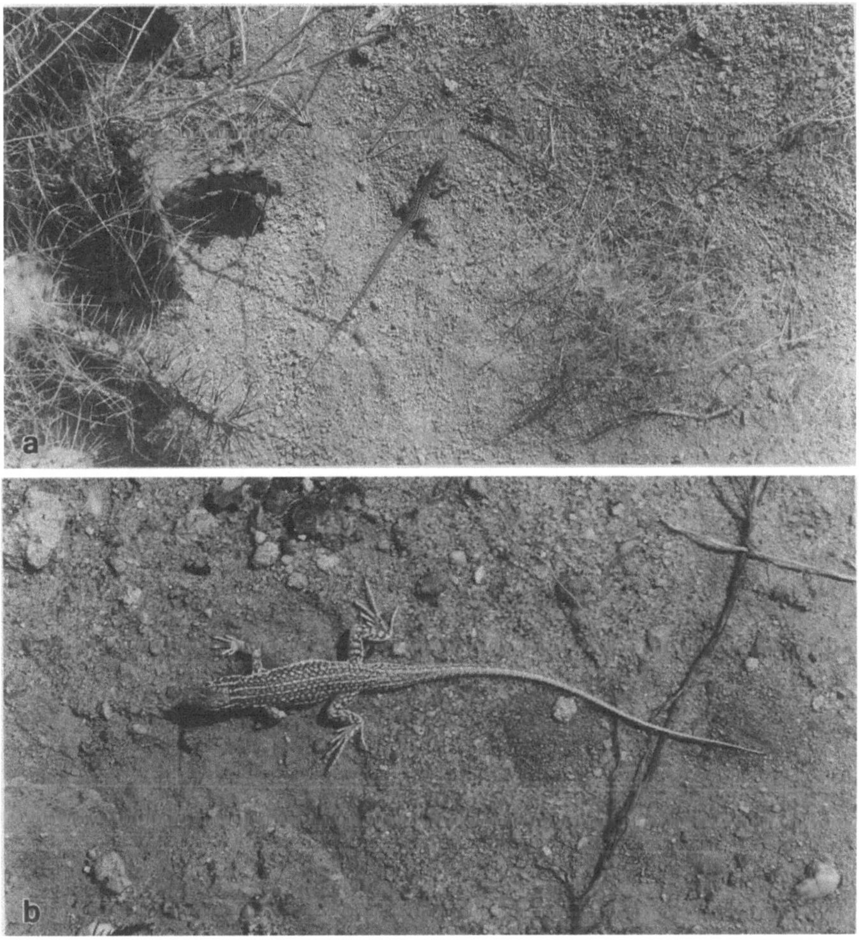

Fig. 9. a *Cnemidophorus* sp. (Teiidae), Chihuahuan desert, New Mexico. **b** *Acanthodoactylus* sp. (Lacertidae), Great Palaearctic desert, Kuwait

desert tortoise *Gopherus* (= *Xerobates*) *agassizii* (Fig. 12a) and the much larger Saharan *Geochelone sulcata* (Fig. 12b), apart from the general similarity common to all Testudinidae, a reptile family adapted for life in dry surroundings.

2.4.2 Parallel Evolution Among Lizard Assemblages

Parallel evolution, convergence and ecological equivalents have been studied in greater detail among lizards than in any other group of reptiles or arthropods. Pianka (1986) examined various aspects of the ecology of lizards in the warm deserts of three continents: the Mojave and Sonaran deserts of western North

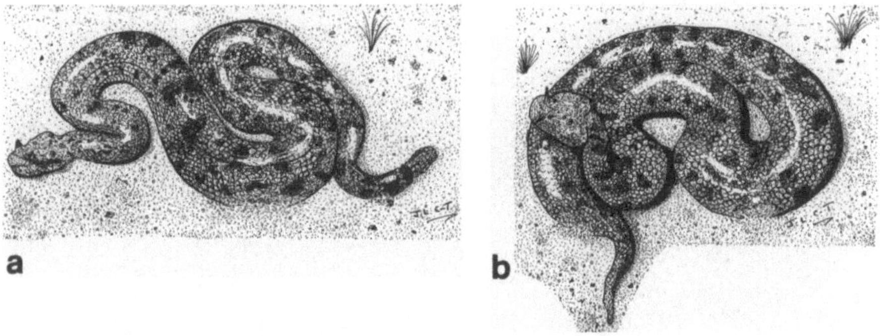

Fig. 10. **a** *Crotalus cerastes* (Crotalinae). **b** *Cerastes cerastes* (Viperinae)

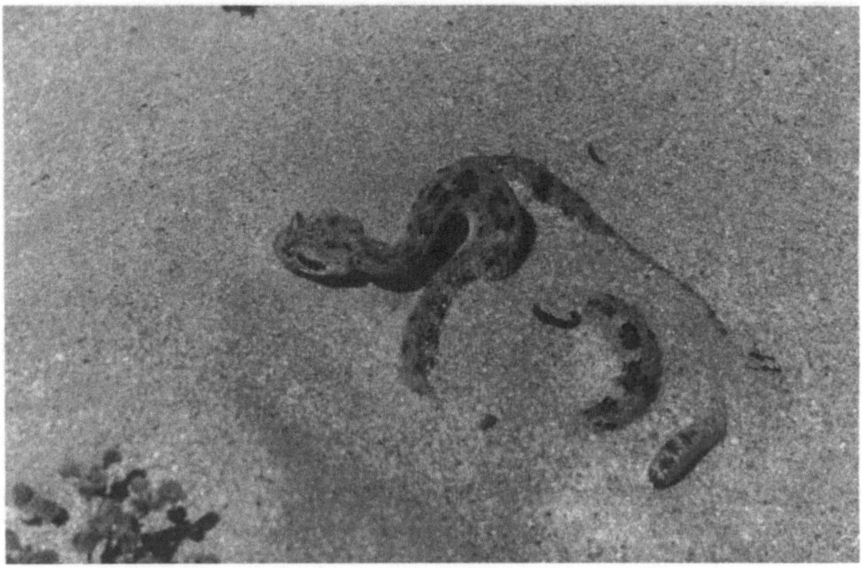

Fig. 11. *Crotalus cerastes*, partly covered with sand during the day, Baja California

America; the Great Victorian desert of western Australia; and the Kalahari of southern Africa. A firm believer in the efficacy of natural selection, he fully expected the Australian deserts to contain a reasonably close replica of the North American desert lizard system, at least on the level of recognizable ecological equivalents. To his surprise, however, he found much greater diversity among Australian than among North American desert lizards, while the Kalahari was intermediate between North America and Australia in this respect. All three systems contain five families, although the identities of these differ between continents.

Fig. 12. **a** *Gopherus* (= *Xerobates*) *agassizii*, Sonoran desert, Arizona. **b** Newly hatched *Geochelone sulcata*, Sahara desert, Sudan

At the generic level of diversity, North America and the Kalahari are similar, but about twice as many lizard genera are recognized in Australia. These intercontinental differences do not stem from diurnal ground-dwelling sit-and-wait foragers, but arise primarily from differences in richness of species that are arboreal, nocturnal, subterranean and, particularly, those that are diurnal, terrestrial and which forage widely. The diversity of insect food resources is highest in North America, and lowest in the Kalahari where termites form a dominant constituent of the diets of lizards. On the other hand, the diversity of microhabitats used by lizards is low in North America and high, but variable, between sites in the Kalahari. The breadths of microhabitat and dietary niches do not differ appreciably between continents, but dietary overlap is highest in the Kalahari due to heavy consumption of termites, lowest in Australia, and intermediate in North America. Microhabitat overlap is essentially all-or-none in North America, but is skewed towards low values in the two southern hemisphere deserts. Overall niche overlap, estimated as the product of dietary, times microhabitat overlap, is lowest in Australia and highest in North America. Here, relatively little guild structure is apparent, although guilds are clearly present in the Kalahari and are conspicuous in Australia. The morphological hypervolume occupied by lizards is largest in Australia, as might be expected, but most compact in the Kalahari where the majority of species are small.

2.4.3 Arthropodan Examples of Convergence

No taxon of arthropods appears to have been investigated in the same way that Pianka (1986) has studied desert lizards. Most desert regions are characterized by the presence of large, black tenebrionid beetles. *Eleodes armata* in North America might, for instance, be regarded as an ecological equivalent of *Blaps inflata* in the northern Sahara. Again, North American harvester ants of the genus *Pogonomyrmex* often construct low, saucer-shaped mounds around the nest entrance (Wallwork 1982) as do some central Australian *Rhytidoponera* spp. (Fig. 13a; Cloudsley-Thompson 1989a) and *Cataglyphis* spp. (Fig. 13b) in the Sahara (Délye 1968). These may serve to protect the nests from sheet flooding, as do the anti-flooding devices of Australian mygalomorph spiders (Main 1976, 1982, Sect. 6.3). The concept of ecological equivalents is almost meaningless in relation to scolopendromorph centipedes, which are all relatively similar in general appearance (Fig. 14a,b), although desert species are usually yellowish while tropical forest species may be brightly coloured.

Convergence is less marked among desert arthropods than among desert reptiles, probably because the former are smaller, more diverse, and therefore occupy less generalized ecological niches. Nevertheless when, for example, comparisons are made between the Kalahari and Namib and the northern Sahara it can be seen that, in contrast to the preponderant species richness of seven genera of Scorpionidae in southern Africa, there is only one species in the north, namely *Scorpio maurus*. This, however, is represented by numerous subspecies, each with differing life-styles, depending upon the texture of the soil. These subspecies are

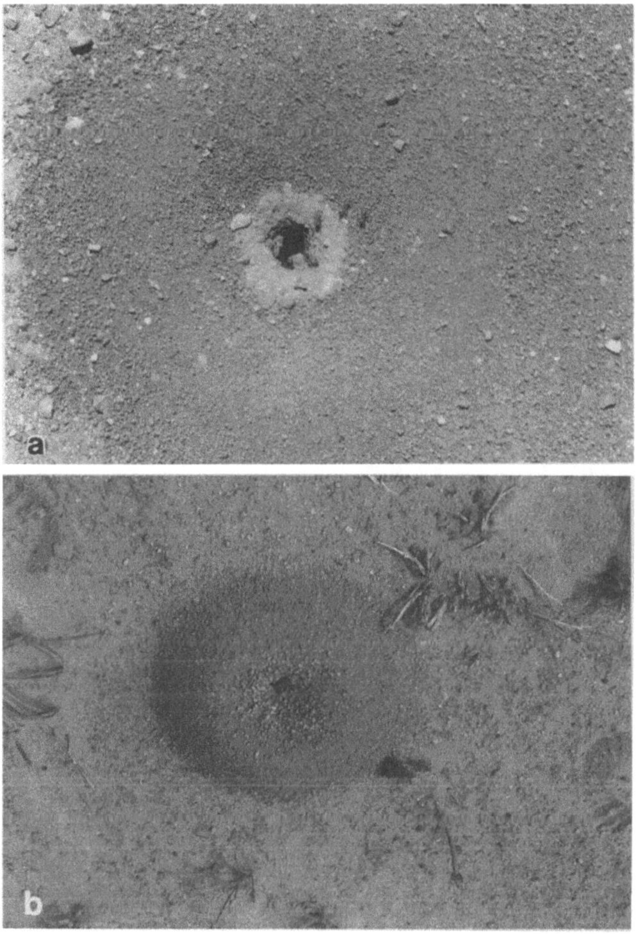

Fig. 13. **a** *Rhytidoponera* sp. (Ponerinae), nest entrance, Simpson desert, Australia. **b** *Cataglyphis* sp. (Camponotinae) nest entrance, Arabian desert, Kuwait

comparable with analogous species of the genus *Opisthophthalmus* in southern Africa. At the same time, the dune-living *O. flavescens* of the Namib occupies a niche comparable with that of the psammophilous *Buthacus arenicola* of the Sahara. Whereas in North Africa woodlice (Isopoda) of the genus *Porcellio* are well adapted to life in sand dunes, where they feed on vegetation and wind-blown detritus, they do not play such an important role in the Namib: here, the major crustacean detritivores of the dunes are fishmoths (Lepasmatidae; Watson 1989; Sect. 7.3.1).

The very fact that one can speak and write about 'desert adaptation' implies that animals in different parts of the world do become positively adapted to the desert environment. Since conditions are broadly similar in all deserts, in that

Fig. 14. a *Scolopendra* sp. (Scolopendrinae), Chihuahuan desert. **b** *Ethmostigmus* sp. (Otostigminae), Australian desert

they are characteristic of a recognizable biome, it would appear that evolutionary convergence must be inevitable. When this engenders morphological and behavioural adaptations that result in relatively unrelated animals becoming similar in appearance, it is probably justified to refer to such species analogues as being 'ecological equivalents'; but this does not by any means imply that they are identical in all respects, as Pianka (1985) has emphasized in the case of reptiles. Indeed, close ecological equivalents are comparatively rare, although conver-

Fig. 15. **a** *Sauromalus* sp. (Iguanidae), Sonoran desert. **b** *Uromastix* sp. (Agamidae), Arabian desert

gence with regard to particular characteristics and behaviour patterns, such as adaptive colouration or sand digging, is common.

Edney (1974) pointed out that there is great structural and physiological diversity among different groups of arthropods, resulting from their very different evolutionary histories, so that the ability to inhabit the desert is not equally open to all. Furthermore, most adult insects are able to fly but other arthropods are not: their adaptations to desert living will differ correspondingly. Desert beetles have very impermeable integuments and tolerate high body temperatures while desert cockroaches live just below the sand surface and have more permeable integuments, but can absorb water vapour. "In fact", he concluded, "there is probably no single aspect in which all desert arthropods differ from all others."

2.5 Energy Flow

Desert ecosystems depend upon the flow of energy through the various trophic levels, and on the cycling of nutrients. The uncertainty of incoming moisture, the principal driving force of the ecosystem, has both direct short-term effects on the consumers of ephemeral vegetation as well as more indirect, long-term effects on larger carnivores, omnivores (including detritivores) and consumers of perennial plants. Most of the ecophysiological research that has so far been carried out on the fauna of deserts has concentrated on physiological adaptations to extremes of heat and lack of water. The biological relationships within the desert biome are still being investigated, and much current research is necessarily descriptive. Nevertheless, the development of computer simulation models has proved to be useful in synthesizing available data, identifying the most important components of the system and generating hypotheses based on what is known (see Goodall et al. 1981).

Ectotherms are at least one order of magnitude more efficient as producers of biomass than birds and mammals of comparable size. A habitat that could sustain only a small population of homeotherms can therefore support a much larger population of arthropods and reptiles. Furthermore, these are inactive for much of the time and at much lower body temperatures than birds and mammals, so their rates of energy intake are perhaps only 1 to 5% of those of homeotherms (Bennett and Nagy 1977).

Crawford (1986) recognized two types of process-functioning patterns among desert invertebrates: trophic level patterns characterizing consumption, and moisture-associated patterns characterizing both consumptive and life-history processes. These can be extended to include reptiles (see below).

2.6 Trophic Level Patterns of Process-Functioning

2.6.1 Herbivores and Detritivores

Non-reserve biomass (Noy-Meir 1973, 1974), which includes stems, leaves, fruits and flowers, as well as dried remains, but not seeds or underground storage organs, form the basis of the 'ecological pyramid' in desert regions. Reserve plant biomass, nectar and pollen are also eaten by many desert insects, themselves the prey of carnivorous insects, arachnids and reptiles. A few desert reptiles feed primarily on plant material. They include tortoises and larger lizards such as *Dipsosaurus, Sauromalus* (Fig. 15a) and *Uromastix* (Fig. 15b) spp.

Many kinds of desert insects forage for nectar and pollen: insect pollination is extremely important to plants in desert regions (Baker and Hurd 1968): Diptera and Hymenoptera are the two orders principally involved. The geographical and ecological distribution of bees (Apoidea) is closely correlated with that of angiosperms. These, and their bee pollinators, are postulated as having evolved in arid regions where non-social bees do most of the pollinating. This may be because the energy levels of long-term nectar and pollen supplies are insufficient to support colonies of social bees. In general, however, the extreme desert environment tends to favour subsocial and social arthropods.

Despite the relatively large amount of subterranean reserve biomass in arid regions, little is known of its consumption by arthropods. Sudden brief flushes or primary production following precipitation may 'swamp' potential grazers and gatherers of nectar and pollen; but dormant stores of concentrated food in the form of roots and seeds are continuously available. Animals exploiting these do not suffer from fluctuations in the availability of food that most other desert dwellers experience. Crawford (1981) has reported that cockroaches, cicadas and the larvae of Coleoptera, Lepidoptera and Diptera are known to forage on the roots or tubers of native desert plants; while seeds are gathered by Bruchidae, other Coleoptera, and especially by granivorous harvester ants (Brown et al. 1979).

In the arid regions of North America, the most specialized seed-eaters belong to the genera *Pogonomyrmex, Pheidole* and *Veromessor* (Myrmicininae). Harvester ants probably evolved from mainly carnivorous ancestors, whose strong mandibles were largely pre-adapted for crushing seeds, and required only slight modification for this purpose. The most significant behavioural adaptation of harvester ants is probably the maintenance within their nests of extensive granaries which can be drawn upon when the standing crop of seed resources has been depleted. The activity schedules of harvester ants are plastic within the bounds dictated by ectothermy and other physical constraints. Activity is predominantly diurnal in colder seasons, and crepuscular or nocturnal during the hot desert summer (Sect. 3.4.1).

Foraging by Chihuahuan desert harvester ants (*Pogonomyrmex* spp.) is affected by soil surface temperature, saturation deficit at the soil surface, light and the availability of food. Intense activity is stimulated by wetting of the soil, and foraging ceases when the granaries are filled (Whitford and Ettershank 1975).

The daily intake and expenditure of energy by the weaver ant *Polyrachis simplex*, which collects honeydew from the cicada *Oxyrrhachis versicolor* in the Negev desert, have been estimated by Gersani and Degen (1988) to be about 50.6 and 1.5-1.7 J respectively (see Sect. 2.8). Energy expenditure is therefore only 3-3.4% of the total intake. The authors concluded that weaver ants have developed an extremely efficient foraging system.

2.6.2 *Carnivores*

Many of the larger desert insects and arachnids, as well as smaller lizards (weighing up to about 50 g), are insectivorous. Foraging by invertebrate predators is often geared to coincide with the availability of the prey, as in the case of the scorpion *Paruroctonus mesaensis* according to Polis (1980a) and presumably, of giant velvet mites (*Dinothrombium* spp.) (Fig. 16), which emerge briefly after the rains that trigger simultaneous emergence of the termites upon which they feed (Cloudsley-Thompson 1962b; Tevis and Newell 1962). Crawford (1986) is of the opinion that foraging, according to the availability of the prey, probably also occurs in desert spiders (Riechert 1978), as well as whipscorpions, scolopendromorph centipedes, etc. (Crawford 1981). According to Riechert (1978), the desert funnel-web spider (*Agelenopsis aperta*) of North America is territorial, and the sizes of the areas occupied by individual spiders are related to both their energy needs and to the accessibility of food within the territories in their particular habitat. Spiders maintain the largest territories where the physical environment is most harsh and the prey unpredictable; while territoriality may break down altogether in areas where prey is abundant and in constant supply throughout the year, and where thermal conditions permit maximum time for its capture. Energy-based territoriality is probably widespread among desert spiders: it helps to explain their distribution. Crawford (1986) believes that invertebrate carnivores in arid regions must make optimal use of foraging space and/or minimal use of foraging time. He cites examples from several different arthropodan taxa.

Small lizards and snakes (weighing up to 50 g) that feed primarily upon arthropods make up the majority of desert reptiles. The most abundant and conspicuous of those in the North American deserts are day-active Iguanidae and Teiidae, according to Brown (1986). Most of the former (e.g. *Sceloporus*, *Uta* and *Phrynosoma* spp.) are intermittently active, sit-and-wait predators, whereas the

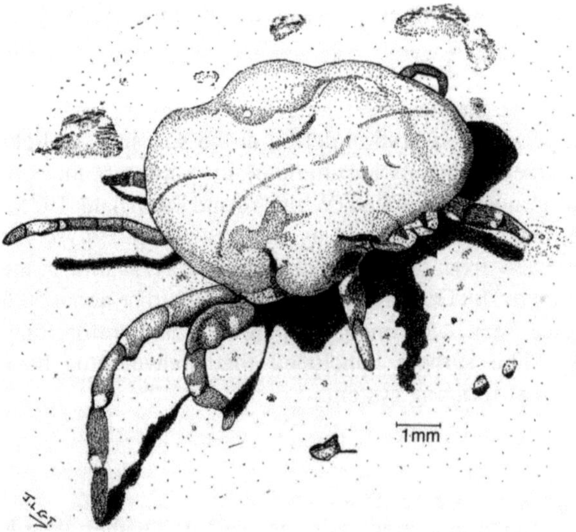

Fig. 16. *Dinothrombium* sp. (Acari), Sahara desert

latter (e.g. *Cnemidophorus* spp.) are more continuously active. In addition to these genera, there are a number of nocturnal insectivorous lizards and snakes.

Larger carnivorous reptiles tend to feed on other vertebrates. In North America they include lizards of the genera *Crotaphytus* and *Heloderma*, and snakes such as *Crotalus, Masticophis, Pituophis* and *Lampropeltis* spp. (mostly in the range of 50-2000 g). Some of the snakes and large lizards are day-active, feeding mainly on smaller lizards, whereas nocturnal snakes prey mostly on small mammals. *Heloderma* spp. eat birds' eggs (Brown 1986). Other desert regions, of course, boast equivalent assemblages of species.

2.6.3 Omnivores

Many detritivores eat living organisms when opportunities occur. Foraging is conditioned mainly by climate and, perhaps less directly, by the quality and availability of food. This applies, for instance, to microarthropods (Whitford et al. 1981), termites and millipedes. Thus, the effects of soil-dwelling desert detritivores on the breakdown of litter and its decomposition are associated to varying degrees with rainfall. The life-history patterns of detritivores can be classified broadly as follows: (1) short lives with relatively rapid responses to changing environments; (2) long lives with relatively few tight linkages to environmental change; and (3) an intermediate set of life histories often complicated by eusociality (Crawford 1979).

Many arid zone lizards are omnivorous in a stricter sense, feeding upon vegetable matter or insects, especially ants, with equal alacrity. *Agama agama* (Cloudsley-Thompson 1981a) and *Uromastix* spp. in Africa are well-known examples. A wide assortment of mineral matter, plant material, fungi and organic debris is found in the guts of invertebrate detritivores (Crawford 1986). Many of them are able to break down the materials ingested with the aid of gut symbionts, some of which are able to fix nitrogen (Crawford and Taylor 1984). This is a useful adaptation in extreme environments, where microbial activity in the soil is low.

Two generalizations apply to dune faunas in arid environments: (1) surface-active arthropod assemblages consist mainly of carnivores and detritivores, and (2) the species compositions of these assemblages differ markedly between day and night and according to season (Holm and Edney 1973). Crawford and Seely (1987) examined the assemblage structure in surface-active arthropods in the Namib desert dunefield, and found that the average species richness of detritivores and carnivores was similar, but individual numbers and total biomass were much greater for detritivores. There were, however, considerable variations between one habitat and another. Although tenebrionid beetles averaged about one-third of all the pitfall-trapped species, their species richness and individual abundance were highly habitat-specific. Diurnally and nocturnally active arthropod species were, on average, equally distributed over day and night.

2.7 Moisture-Associated Patterns of Process-Functioning

Some desert arthropods are short-lived and/or primary consumers, especially those that feed on ephemeral vegetation. They have *r*-selected life cycles that are closely correlated with rainfall. The pulses of precipitation that 'drive' primary production (Noy-Meir 1974) are more important than either photoperiodic changes in day length or temperature in synchronizing seasonal activity (Cloudsley-Thompson and Idris 1964). The second moisture-response group includes comparatively long-lived, K-selected species, whose proximal responses to moisture are of less relevance than general developmental strategies attuned to a variety of stimuli (Crawford 1986). Desert reptiles, like arthropods, occupy various positions along the *r*-K continuum of selection.

Three unrelated patterns of invertebrate response to aridity are cited by Crawford (1986): (1) invertebrate populations tend to exhibit peak numbers early in the dry season; (2) a relative increase in body size is favoured in arid environments (Remmert 1981); and (3) there is a tendency towards social behaviour. This response will be discussed below (Sect. 3.5). It should be emphasized, however, that these generalizations are little more than trends, and numerous exceptions to them can be found.

Although conservation of body water is a critical factor in the survival of both arthropods and reptiles in arid environments (Cloudsley-Thompson 1964a, 1971, 1972a, 1975; Mayhew 1968; Ahearn and Hadley 1969; Ahearn 1970b,c; Edney 1974, 1977; Crawford 1981; Bradshaw 1986; etc.), Willmer (1982) has suggested that water conservation may, in certain circumstances, be a secondary consideration compared with an insect's selection of an appropriate thermal regime. This assertion was based in part on Hamilton's (1971) study of *Cardiosus* spp. (Tenebrionidae) in the Namib desert, and on the work of Dreisig (1980) on *Cicindela hybrida* (Cicindelidae) in sandy habitats of Denmark, which showed that these insects select microclimates and activity times based on thermal regimes rather than on humidity. Hamilton (1971) termed one aspect of this behaviour 'maxithermy': his hypothesis has been rejected by Heinrich (1977), Larmuth (1979) and the present author (Cloudsley-Thompson 1979a), although it has received considerable support from other workers (e.g. Seely 1985) and will be discussed further below (Sect. 4.3.2; see Hadley 1979a,b).

The conclusions of Hamilton (1971) and Dreisig (1980) are not incompatible, however, with the claim, stated above, that both arthropods and reptiles are pre-adapted to the desert environment insofar as water conservation is concerned. Indeed, Dreisig (1985) has presented a time-budget model of thermoregulatory sun-basking, based upon physical laws, that applies equally to insects and lizards.

2.8 Field Metabolic Rates

In addition to that of harvester ants (Sect. 2.6.1), the daily cost of activity and expenditure of energy in the field have been studied in tenebrionid beetles (Parmenter et al. 1989), scorpions (King and Hadley 1979) and desert lizards

(Anderson and Karasov 1981). In the Mojave desert, the tenebrionid *Eleodes armata* had standard metabolic rates that were higher than those of *Cryptoglossa verrucosa* over the range 10-30 °C, while its Q_{10} was only 1.4 compared with 2.6 for *C. verrucosa*. Field metabolic rates, however, were significantly different between the two species in spring and winter, while activity costs (total field metabolism/resting field metabolism) ranged between 2 or 3 for both species over all seasons. *E. armata* was estimated to ingest about ten times the dry mass ingested by *C. verrucosa* during the year. *C. verrucosa* adults, with their lower metabolic rates and daily energy expenditure, are at an advantage in regions with low primary productivity, which is consistent with the hypothesis that this species evolved in an arid environment. In contrast, *E. armata* has probably been able to invade the desert partly as a result of exploitation of the seasonal availability of food (P.D. Cooper, pers. commun. 1990).

Four species of *Eleodes* inhabiting different microhabitats in an arid, sagebrush-steppe system of south-west Wyoming (USA) were found by Parmenter et al. (1989) to have significantly different metabolic rates. These, however, like rates of water loss, did not fit predicted distributions although, by virtue of their cooler, preferred microhabitat selection, shrub-dwelling species could potentially retain more water and metabolize less energy than the species inhabiting warmer microhabitats. In the case of free-roaming scorpions (*Hadrurus arizonensis*), King and Hadley (1979) calculated that respiration rates were two or three times higher than those determined under laboratory conditions at comparable temperatures.

The widely foraging lizard *Cnemidophorus tigris* has significantly higher rates of field energy metabolism (210 J g^{-1} day^{-1}) and water flux than the smaller sit-and-wait predator *Callisaurus draconoides* (136 J g^{-1} day^{-1}) in the Colorado desert of California (Anderson and Karasov 1981). Again, Nagy et al. (1984) found that the energy costs associated with active foraging by the lacertid lizard *Eremias lugubris* (800 J g^{-1} day^{-1}) were significantly higher than those of the sympatric sit-and-wait predator *E. lineoocellata*. Resting metabolic rates were similar in both species, but field metabolic rates, on a 24-h basis, were 3.1 times greater than resting rates in *E. lugubris*, and 2.2 times greater in *E. lineoocellata*. *E. lugubris* made a greater profit on its investment (46 vs 36% each day) and grew nearly twice as fast as *E. lineoocellata*. Once again, the metabolic rates of active individuals were up to three times greater than those of inactive animals, yet a widely foraging mode of life seems to provide greater rewards that a sit-and-wait strategy in arid regions. This conclusion is supported by the greater activity and more pronounced circadian locomotory rhythms of desert- than of forest-inhabiting predatory arthropods (Cloudsley-Thompson 1981b; Sect. 3.3.1).

2.9 Conclusions

Processes underlying the dynamics of desert ecosystems evidently reflect the extremes of the physical environment, but they are not well understood in detail. In general, however, it is clear that the most important components of desert food chains include: (1) detritivores and herbivorous insects that feed upon the 'non-reserve' plant biomass; (2) seed-eaters and nectar-pollen foragers that harvest 'reserve' plant biomass; (3) insectivorous arachnids and smaller reptiles; (4) larger carnivorous reptiles that prey upon the latter. (Birds and mammals have been excluded from consideration here.) The principal driving force of desert ecosystems is the input of moisture in the form of rain or fog. At the same time, water conservation may, in some instances, be a secondary consideration compared with the selection of an appropriate thermal regime. This is in accordance with the hypothesis that arthropods and reptiles are largely pre-adapted to life in arid regions insofar as water conservation is concerned. An ectothermic mode of life is beneficial in deserts because it can be supported by a low rate of energy intake. Nevertheless, active foraging by carnivores pays better than a sit-and-wait policy when food is scarce.

3 Avoidance of Environmental Extremes

3.1 Introduction

A basic problem facing all desert organisms is the maintenance of an equable temperature without using an excessive amount of water for evaporate cooling. In smaller animals such as arthropods and most reptiles, whose surface to volume ratio is correspondingly high, this problem, always present on land, becomes even more acute (Cloudsley-Thompson 1988a). Ambient temperature is important to living organisms because many of the chemical, physiological and physical processes upon which life depends are temperature-dependent. Extremes of temperature — either too high or too low — can be lethal, and animals subjected to wide changes in the temperature of the environment must, if they are to survive, be able either to regulate or to control their body temperatures adequately. Small terrestrial animals exchange heat with the environment by conduction, convection, radiation and evaporation. In the case of most day-active arthropods and reptiles, heat is gained mainly by radiation and lost chiefly through convection (Willmer 1982). The obvious way of avoiding radiation is to keep in the shade.

Arthropods and reptiles are ectothermal or brachymetabolic: their metabolic rates are low, as we have seen, and most of their body heat is obtained from the environment. When their temperatures reach or exceed the optimum, however, they seek cooler microhabitats. Thus, desert forms tend to avoid the heat of the summer days by sheltering in cool, shady retreats from which they emerge only at dusk and dawn, or during the night. Emergence is timed by their circadian rhythms or biological 'clocks', and thermoregulation is reinforced by behavioural responses. Although physiological thermoregulation takes place to some extent, it is far less effective in ectotherms than it is in homeotherms because metabolic rates are so much lower. Before discussing thermoregulation, however, we should first consider the microclimates in which desert animals take shelter from the extreme daytime heat and nighttime cold of the desert.

3.2 Desert Microclimates

Although ectotherms are largely dependent upon secluding themselves in favourable microenvironments in order to maintain a degree of thermal homeostasis, surprisingly little information is available about physical conditions in the burrows and other retreats used by small desert animals. Among the first direct microclimatic measurements of temperature to be made were those of C.B. Williams who resided in Egypt between 1921 and 1927. His work was published in *Ministry of Agriculture of Egypt Technical and Scientific Bulletins*, and summarized 30 years later (Williams 1954). It demonstrates that, within a distance of relatively few metres, there may be a very wide range of temperatures available in desert country from among which an animal can choose with little energy expenditure. By burrowing, it can avoid the extreme heat of the day and, by leaving the burrow at night, it can even escape the peak temperature below since, 30 cm or so beneath the surface, there is a time lag of about 12 h in the onset of maximum temperature.

These conclusions were confirmed by measurements made in Tunisia in April 1954 (Cloudsley-Thompson 1956), while further observations, made in the Red Sea hills and coastal plain in 1960, showed that, given a very small reduction in conditions of extreme heat or dryness, an ecological chain can depend on even such apparently unrewarding material as dry vegetable matter (Sect. 2.6.1). Animal distribution tends to be influenced by extremes rather than by means. In one range of *jebels* or hills, a maximum surface temperature of 83.5 °C was recorded at 13.00 h. The only animal to be seen was a solitary grasshopper but, 4 h later, when the sand temperature had dropped to about 40 °C, some ants ventured out. Surprisingly, the colour of the sand made comparatively little difference to its temperature — only 1-2 °C. Rock surface temperatures, however, were influenced to a greater extent by colour; differences between light-coloured rocks and others coated with dark desert varnish being about 3 °C. Low humidities were recorded even among the roots of the grasses where most of the arthropods were found, and the primary biological advantage of this habitat was apparently in the reduction in temperature afforded by the shade (Cloudsley-Thompson 1962a).

Near Biskra, Algeria, the ground temperature in July fluctuates between 30 °C at night and 70 °C at noon, while air humidities vary between 50 and 0% RH (Coenen-Stass 1989). Similar microclimatic conditions have been recorded in northern Africa among others by Messina et al. (1977) and Heatwole and Muir (1979). Below a depth of about 30 cm, soil temperature remains relatively constant around 35-40 °C throughout the day, soil moisture is about 10%, and the air in the burrows of the desert woodlouse (*Hemilepistus reaumuri*; Fig. 28a) is continually saturated. It remains saturated when the soil moisture is only 10% RH and its water content 6-7%; while the aggregation of an entire woodlouse family induces high air humidity within the narrow burrow when the water content of the soil is even lower (Coenen-Stass 1989; Sect. 3.5).

Undoubtedly, temperatures within burrows are very similar to those in sand or soil at the same depth. Edney et al. (1974) recorded maximum and minimum temperatures on each of 14 days distributed throughout a year on the sand surface and at a depth of 5 cm near Palm Springs, California. The maximum range was 48 °C at the surface, but only 23 °C at -5 cm where the daily range was always less than that at the surface. It is characteristic of desert soils that a large daily swing of temperature on the surface, coupled with progressively smaller swings at greater depths, leads to a daily thermal inversion (Cloudsley-Thompson 1956; Déley 1968; Hamilton 1971; Holm and Edney 1973; Edney et al. 1974).

The thermal significance of burrowing deeply is well illustrated by the work of Hadley (1970b) who measured the daily temperature sequences at various levels above and below an open desert surface in Arizona, as well as the burrow temperatures experienced by free-ranging scorpions (*Hadrurus arizonensis*; Fig. 17). Although the soil surface temperature fluctuated 46 °C over a 24-h period, the burrow temperature variations, some 20 cm below, did not exceed 10 °C. Comparable results have been obtained by Shorthouse and Marples (1980) in their study of burrowing behaviour in the Australian arid zone scorpion *Urodacus yaschenkoi*. Likewise, Seymour and Vinegar (1973) found that daily temperatures in the burrows of tarantulas (*Aphonopelma* sp.; Fig. 75) varied only by some 3 °C at a depth of 22 cm while the solid surface outside the burrows fluctuated over 24.5 °C. At a depth of 20-30 cm, where the white lady spider *Leucorchestris arenicola* (Heteropodidae) digs its burrows in the Namib dunes (Henschel 1991),

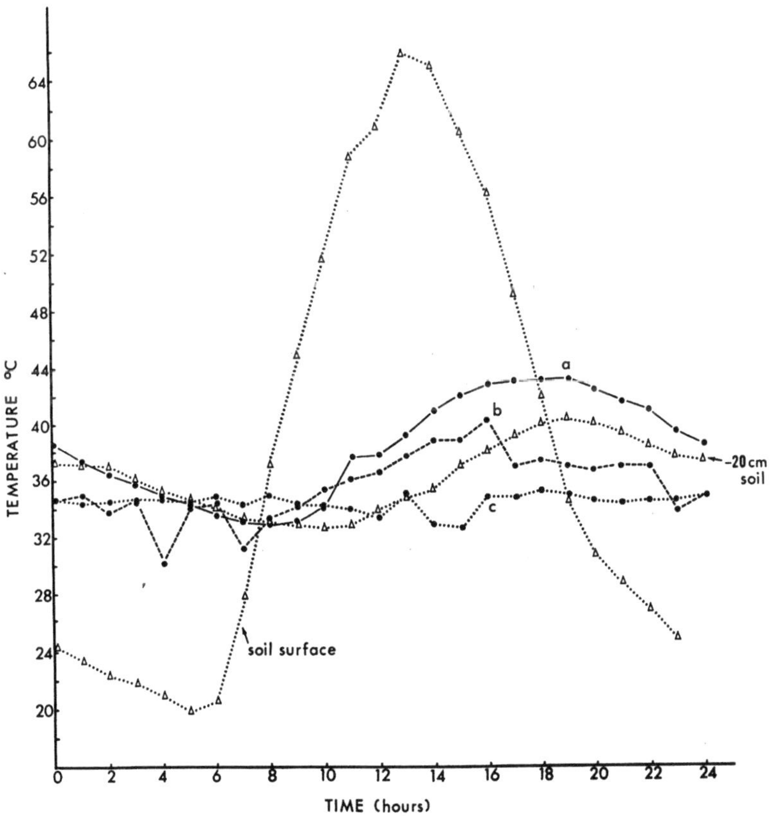

Fig. 17. Burrow temperatures experienced by three scorpions (*a,b,c*) at various depths over a 24-h period. Surface and -20 cm soil temperatures during the same period are included. All scorpions remained in the burrows throughout the 24 h (Hadley 1970a)

sand temperatures fluctuate little on a daily basis but dip seasonally from ca. 30 °C in summer to ca. 20 °C in winter (Lancaster et al. 1984). Morphological adaptations for digging burrows will be discussed below (Sect. 7.2; Fig. 18a,b).

Many desert arthropods and reptiles possess only weak powers of burrowing, but they are able to excavate or extend the spaces beneath rocks and stones, the microclimatic regime beneath which is of paramount importance to their survival. In an experimental study carried out in Morocco, Larmuth (1978) demonstrated that the temperature reached beneath a stone is dependent on its thickness and

Fig. 18. a Scorpion burrow, Arizona. **b** Lizard burrow, Kuwait

colour, not on its horizontal dimensions. Colour made a considerable difference (13.1 °C) in the temperature beneath a stone 2 cm thick but, with increasing thickness, colour resulted in progressively smaller differences until, at 10 cm, the temperature was only 5.8 °C less below a white marble slab than below a grey one. Thickness of a stone also affected the temperature registered beneath it. For example, grey stones shared a difference of 10.1 °C between the temperatures below thicknesses of 2 and 10 cm. In contrast, a change of only 2.8 °C was observed under white stones with a similar increment in thickness. The surface temperatures of the stones were not measured.

The measurements made in the Red Sea hills, referred to above, suggest, however, that a considerable amount of heat would have been radiated to the environment and that there must have been a slow buildup in the cavity beneath the stones. The base of the cavity beneath a stone was cooler than the underside of the stone: the differences were approximately proportional to the depth of the cavity. Furthermore, differential heating of the side of a stone resulted in thermal gradients beneath it depending upon the ratio of the horizontal to vertical surface.

Sand and rock provide unique thermal niches in hot deserts. On the whole, rock temperatures fluctuate less than sand (George 1986). Peel (1974) recorded 79.3 °C on varnished sandstone in mid-August in Tibesti at a midday air temperature of 47 °C, but the differences between different types of rock were minimal. Peel found basalt to be a few degrees hotter than varnished sandstone; while darker rock was generally slightly hotter than pale rock. Rock surface tends to provide a more equable temperature than sand, and provides convenient pockets of shade. George (1986) found that rock lizards have a broader thermal niche and, on the whole, are not active at such high temperatures as are sand-dwelling lizards.

Little information is available regarding the protection afforded by vegetation (Cloudsley-Thompson 1962a). In a comparison between microclimates during the hottest part of the day, Larmuth (1979) found that, in Morocco, tufts of grasses (*Lygeum spartum*) offered a more protracted humidity regime than bushes (*Salsola gastola*) but that, at similar temperatures, refuges with higher humidity were preferred by insects, e.g. the beetles *Ammocteonous hieroglyphicus, Pimelia anomalus* and *Oterophloeus picipes*. When the soil was shaded by a common bunch grass, which influenced short- and longwave radiation, as well as wind speed profiles, simulated maximum soil surface temperatures decreased about -2 °C for every 10% increase in shading, indicating that plants can have a substantial effect on the local microclimate (Nobel and Geller 1987).

3.3 Circadian Locomotory Rhythms

Emergence from burrows and retreats, with their equable microclimates, is controlled both by responses to environmental factors and by biological clocks which synchronize the activities of burrow-dwellers with changing daily conditions in the outside world. Most arthropods and reptiles show circadian rhythms of locomotory activity, but those of desert species tend to be especially well marked. Such subjective impressions are difficult to prove experimentally but, as mentioned above (Sect. 2.8), there is evidence that desert centipedes, scorpions and mygalomorph spiders are not only more strictly nocturnal than temperate and tropical forest species, but they are generally much more active (Warburg and Ben Horin 1979; Constantinou 1980; Cloudsley-Thompson 1981b). Scorpions and Solifugae are unusually tolerant of high temperature (Sect. 4.3.1) and have very low transpiration rates (Sect. 5.2.1.1). For this reason, it seems probable that the avoidance of predators may be of greater significance than thermal physiological requirements in their night-active behaviour (Cloudsley-Thompson 1961a,b; Sect. 7.2). At the same time, of course, the activity of a predator must be timed to coincide with that of its prey.

3.3.1 Arthropods

Rhythmic locomotory activity in desert Arthropoda has been described in many books and reviews, e.g. in woodlice (Isopoda: Oniscidea) (Cloudsley-Thompson 1977c; Sutton and Holdich 1984), myriapods (Chilopoda and Diplopoda) (Crawford 1981) and Arachnida (Cloudsley-Thompson 1975, 1978), including scorpions (Hadley 1974; Warburg and Polis 1990), Solifugae (Cloudsley-Thompson 1977b) and desert spiders (Cloudsley-Thompson 1983), as well as in numerous insect families, especially Tenebrionidae (Cloudsley-Thompson 1961a,b, 1975; Crawford 1981; Applin et al. 1987).

The physiological basis of circadian rhythms in desert arthropods has been studied in considerable detail (reviewed by Applin et al. 1987). Numerous investigations have been concerned with establishing the physiological nature of the biological clocks rather than with their employment as a facet of desert adaptation: this aspect of circadian rhythms will not be discussed in the present book.

Most desert animals show extremely well-marked circadian rhythms of locomotory activity both in nature and under experimental conditions. At least seven species of Lepismatidae (Thysanura) occur sympatrically in parts of the dune sea of the Namib desert, where they occupy different areas of the habitat (Sect. 7.3.1) and are active at different times of day. According to Watson (1989), variation in spatial use may be explained as a response to the physical environment. Some species are found mainly on slip faces, others in clumps of vegetation, where the biomass of detritus is high; while bare sand slopes tend to be avoided. Variations in patterns of temporal activity, whether day-active, crepuscular, nocturnal, or dawn-active, can be explained as responses to specific environmental conditions and may be related to the species' physiological adaptations that permit activity under different weather regimes. Not surprisingly, there is some spatial as well as temporal overlap.

An exception to the rule that desert animals tend to be markedly rhythmic is afforded by the desert locust *Schistocerca gregaria*. Although activity of the locust is greater in constant light than in darkness, the periodicity is labile. A marked rhythm of locomotory activity and rapid synchronization with environmental factors might, however, be a disadvantage to a desert locust experiencing the inconstant environment of a semi-arid region in which any temporary advantages must be exploited immediately. For instance, the availability of food, or of moist sand for oviposition, is so ephemeral that locusts could not afford to wait until dawn or dusk before eating or laying their eggs. In keeping with this, a circadian rhythm that was easily and irreversibly synchronized by changes in temperature or light intensity would be disadvantageous. The lability of locust activity rhythms and the weak influence of their environmental synchronizers should both be regarded as adaptations to the desert environment (Cloudsley-Thompson 1980a).

A comparative study of the reactions to temperature of the Saharan tenebrionid beetles, *Adesmia antiqua* and *Pimelia grandis*, disclosed that the preferred temperature of the day-active *A. antiqua* is 20-40 °C, that of the nocturnal *P. grandis* 20-33 °C (El Rayah 1970a). When the day-active *Adesmia cancellata* of Kuwait was compared with the nocturnal *Trachyderma hispida* which, like *T. philistina* (Fig. 19), occupies the same habitat, Abushama and Al-Salameen (1989) found that, contrary to what might have been expected, *T. hispida* exhibited a preference for a wider range of temperatures and lost less water by transpiration than did *A. cancellata* under the same experimental

Fig. 19. Photonegative response of *Trachyderma philistina*, Kuwait

conditions. This may suggest that *A. cancellata* thermoregulates by adapting to higher rates of transpiration than *T. hispida*. On the other hand, *A. cancellata* can lift its body from the hot desert surface with its long legs and, when venturing into the open, run fast from one patch of shade to another — as does *Onymacris plana* in the Namib dunes. Bolwig (1957) concluded that day-active tenebrionid beetles are adapted to tolerate high temperatures for short periods and have a mechanism that makes it possible for them to maintain a body temperature of 40 °C for a period long enough to cross exposed sand from one place of shelter to another. Whereas, in hot deserts, the majority of arthropods, including many species of desert beetles especially Tenebrionidae (Fig. 19) and Carabidae (Fig. 20), are nocturnal (Cloudsley-Thompson and Chadwick 1964), in cool, coastal deserts there are often more diurnal than nocturnal species. This may well be correlated with the ability of the day-active forms to thermoregulate efficiently

Fig. 20. a *Thermophilum venator*, Tunisia. **b** *T. duodecimpunctata*, Kuwait

by shuttling between sunshine and shade, or by burrowing in the sand (Sect. 4.2.2.1). The Atacama and Namib deserts, moreover, are relatively cold at night which, for metabolic reasons, puts a premium upon activity during the daytime. Over half the species of Tenebrionidae in the Namib, for instance, are day-active (Koch 1961; Cloudsley-Thompson 1990; Crawford et al. 1991) and there are even some diurnal Solifugae, such as *Metasolpuga picta*, which show thermoregulatory behaviour, not previously noted in this order and associated with diurnal habits (Wharton 1987).

Some confusion has been caused by the fact that although aktograph experiments in the laboratory invariably show a marked circadian rhythm in scorpions (Cloudsley-Thompson 1956, 1981b; Warburg and Ben Horin 1979; Constantinou 1980; etc.), fieldwork by Polis (1980a) has indicated that, although emergence from their burrows at dusk is correlated with decreasing levels of ambient light intensity, *Paruroctonus mesaensis* in California does not emerge every night from its burrow. The explanation may well be that, although their circadian rhythms stimulate the scorpions so that they come to the mouths of their burrows, they do not emerge unless they require food and climatic conditions outside are favourable. Such appears to be the case with *Urodacus yaschenkoi* according to Shorthouse and Marples (1980). This conclusion is also supported by the elegant experimental investigations of Fleissner and Fleissner (1985) on the circadian rhythm of pigment migration in the median eyes of *Androctonus australis*. Even though this strictly nocturnal species is not normally exposed to daylight, its hypersensitive rhabdomes are protected, as though by sunglasses, by the outward movement of screening pigment. In the event of untoward exposure during the day by the accidental removal of the rock or boulder beneath which they are sheltering, the scorpions are pre-adapted for any such eventuality and are not blinded by the intensity of light suddenly experienced. There are also age-specific differences in seasonal and nightly patterns of surface activity which minimize cannibalism and competition for food among scorpions.

3.3.2 Reptiles

As in arthropods, the daily activities of most reptiles are controlled by endogenous circadian rhythms (Cloudsley-Thompson 1971, 1988b; Heatwole 1976; Pianka 1986). These, too, are synchronized or entrained by fluctuating cycles of light and/or temperature, the relative effects of which vary among different species. Such external factors may inhibit activity at extreme levels (Cloudsley-Thompson 1971; Heatwole 1976). Humphreys (1978) emphasized the similarity in thermoregulatory behaviour between the Australian burrow-inhabiting wolf spider *Geolycosa godeffroyi* and desert lizards.

The time of day at which a reptile is active, plays an important part in its adaptation to life in the desert. As Huey (1982) pointed out, the daily and seasonal activity patterns of reptiles are strongly influenced by their thermal biology, i.e. species with the highest preferred temperatures are typically active at the hottest times of the day and year. Nevertheless, the relationship is extremely complex:

limits to potential activity are set by both the biophysical characteristics of the species and also by the environmental influences it experiences.

The daily rhythm of activity and rest has been studied in numerous desert reptiles, and is closely associated with thermoregulatory behaviour. Although the principles are similar in both taxa, far more is known about reptiles than about arthropods in this context. Species of reptiles that live in the same habitat may co-exist because they differ in their behaviour, in their food, in their shelters and retreats, and in the times of day at which they are active. Temporal separation is maximal between night and day while, during these periods, different thermal responses also separate species and thereby facilitate their co-existence. The temperatures at which a reptile is active define its thermal niche because, in hot deserts especially, surface temperatures of the sand and rock in shade or under the blazing sun may differ markedly.

The lizards of Arabia, like those of other desert regions, can therefore be separated ecologically into two major groups — diurnal and nocturnal — the latter mainly comprising geckos. Within these two groupings, species differ principally in their microhabitats (particularly such features as distance from vegetation and the softness of the ground), hunting techniques and the sizes of their mainly arthropod prey (Arnold 1984b).

The North American collared lizard *Crotaphytus reticulatus* follows a typical diurnal pattern which begins with early morning basking, followed by searching for prey, intensive sunning activity, retreat from extreme temperatures and re-emergence for afternoon basking and limited foraging until dark (Montanucci 1971). In contrast, the regal horned lizard *Phrynosoma solare* is active mostly at dawn between 06.00 and 09.00 h in Arizona (Parker 1971).

Unlike the Mojave desert chuckwalla *Sauromalus obesus* (Fig. 15a), the larger *S. hispidus*, endemic to several islands in the Gulf of California, remains active throughout the summer. In winter and spring its rhythm is unimodal, but it becomes bimodal during the summer months. These larger chuckwallas are active for less than 2 h after emergence during the hotter months of July and August, however, and they emerge and seek shaded retreats at earlier times in spring and summer than in the cooler periods of the year (Smits 1985).

As in the case of most desert agamid lizards, *Agama persica* of the Arabian peninsula is day-active, and shows a marked endogenous rhythm which is synchronized by changes in light intensity. The African rainbow lizard *A. agama* is likewise day-active (Fig. 21a). It roosts in vegetation (Fig. 21b) and on trees or buildings at night (Cloudsley-Thompson 1981a). By means of aktograph experiments it has been shown to possess a typical endogenous circadian rhythm of locomotory activity such as has been found in the case of all reptiles investigated except for a few burrowing forms (see below). Chong et al. (1973) have shown that the threshold for panting of the Australian *Amphibolurus muricatus* varies with time of day: daytime thresholds are higher than nocturnal ones. This has obvious adaptive advantages: the threshold is highest during the day, consequently respiratory water loss is kept at a minimum. Firth and Heatwole (1976) found that the rhythm is abolished by parietalectomy and shielding the eyes.

Fig. 21. **a** *Agama agama*, a day-active lizard, Nigeria. **b** *A. agama* roosting at night, Nigeria

Most geckos, such as *Cyrtodactylus scaber* (Fig. 22), are nocturnal but a few, such as the Asian stone gecko *Bunopus tuberculatus* (Fig. 23), are day-active. Variations occur even within the same genus, however. For example, within the Great Palaearctic Desert, *Hemidactylus brookii* (Fig. 24) is strictly nocturnal, as is the Mexican gecko *Hemidactylus frenatus*, whilst *H. turcicus* (Fig. 25) appears

Fig. 22. *Cyrtodactylus scaber*, a nocturnal gecko, Qatar

Fig. 23. *Bunopus tuberculatus*, a day-active gecko, Kuwait

Fig. 24. *Hemidactylus brookii*, a nocturnal gecko, Nigeria

Fig. 25. *H. turcicus*, a crepuscular gecko, Oman

to be crepuscular, sometimes even emerging in full daylight. Activity shows a close correlation with air temperature in *H. frenatus* according to Marcellini (1976). Light intensity, however, determines the initiation and cessation of the diel activity cycle. The animals emerge at dusk: their numbers increase to a maximum around midnight and remain high until they return at dawn to their daytime retreats.

The times of activity of three subspecies of the gecko *Ptyodactylus hasselquistii* in Israel and Sinai are related to geographical distribution. *P.h. puiseuxi* has the northernmost distribution in upper and eastern Galilee, *P.h. hasselquistii* occurs at Elat, the surrounding mountains, and in southern Sinai, while *P.h. guttatus* is common throughout Israel and northern and central Sinai. *P.h. puiseuxi* is the most diurnal, *P.h. guttatus* less so, and *P.h. hasselquistii* the most nocturnal. The activity patterns shown by these three geographical forms can be explained by the fact that northern Israel is colder than in the south (Frankenberg 1976, 1979, 1981). The difference in diurnality between *P.h. puiseuxi* and *P.h. guttatus* is reflected by the sizes of their eyes, those of the latter being much larger (Werner 1969). The degree of diurnality of Israeli geckos is thus correlated with their distribution, those living in arid regions are more nocturnal than species which inhabit more humid areas (Frankenberg 1978).

The granite night lizard *Xantusia henshawi* (Xantusiidae) of North America, in contrast, is primarily day-active, especially in winter. In the field, the period of apparent activity coincides with relatively high body temperatures, suggesting behavioural thermoregulation during the day (Mautz and Case 1974). The day-active, fringe-toed sand lizard *Acanthodactylus schmidti* (Lacertidae; Fig. 9b) possesses an endogenous circadian rhythm of activity and rest which is synchronized by both light-dark cycles and by regularly fluctuating temperature. In response, the animals emerge from their burrows about the same time each morning and retire to rest about the same time each night, regardless of the ambient temperatures (Fig. 26; Constantinou and Cloudsley-Thompson 1985). *A. schmidti* resembles the Sudanese *Mabuya quinquetaeniata* (Scincidae) with regards to the synchronization of its biological clock by light-dark cycles (Cloudsley-Thompson 1965a), but its response to temperature fluctuations is greater. These differences could be correlated with the different desert environments inhabited by the two species: the winters in Kuwait are far colder than they are in the Sudan, so there may be a selective advantage to *A. schmidti* in possessing a circadian activity rhythm that is more dependent upon temperature than is that of *M. quinquetaeniata*.

Although most skinks are day-active, a few little-known Australian species of the genera *Sphenomorphus* and *Egernia* are nocturnal and spend the daylight hours in elaborate tunnel systems which they dig in sandy parts of the Great Victoria desert of Australia. They are also markedly seasonal and, during winter, hibernate in a blocked-off side tunnel, while the remainder of their burrow system falls into disrepair (Pianka and Giles 1982). Activity above ground peaks in the late morning and early afternoon in whiptail lizards, *Cnemidophorus* spp. (Teiidae; Fig. 9a). As the number of lizards above ground increases, in *C. sexlineatus*, so do the levels of aggression (Leuck 1982). The Australian skink

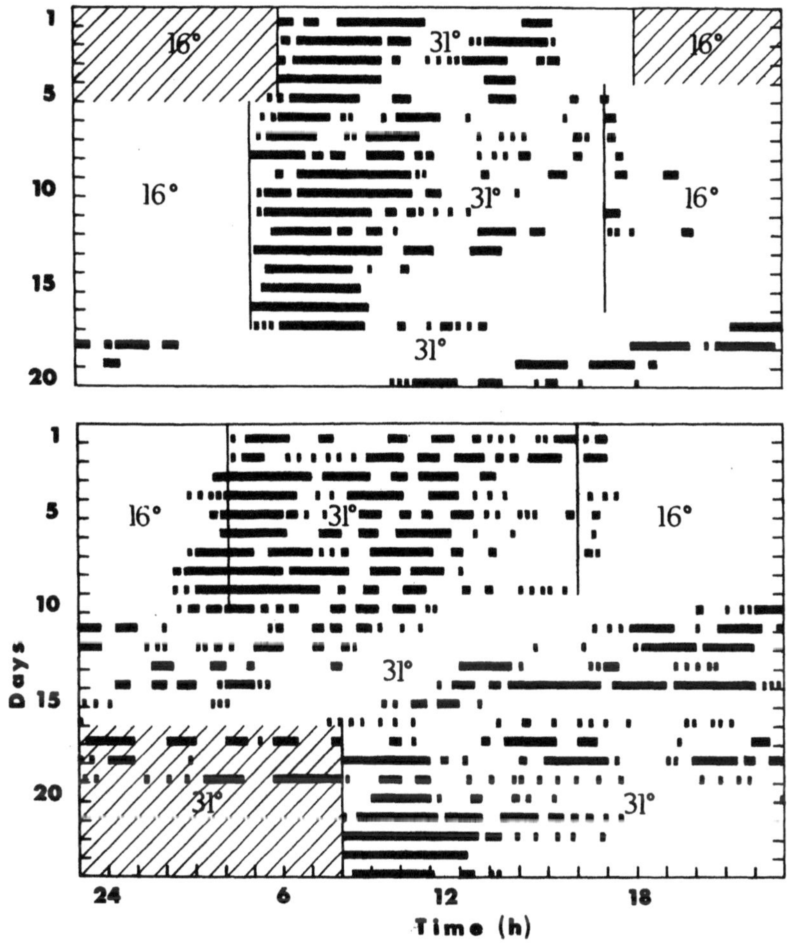

Fig. 26. Activity of two individual *A. schmidti* in aktograph apparatus under different lighting and thermal regimes. *Above* In L:D/12:12 with fluctuating temperature (days 1-5); LL (constant light) with fluctuating temperature (12 h at 16 °C, 12 h at 31 °C; days 6-16); LL at 31 °C (days 17-20). *Below* In LL with fluctuating temperature (days 1-10); LL at 31 °C (days 11-16); L:D/18:8 at 31 °C (days 17-24) (Constantinou and Cloudsley-Thompson 1985)

Tiliqua scincoides shows no diel variation in basic heart rate, which is inversely related to body weight, but resting lizards have lower heart rates in constant darkness than they do when maintained in constant light (Firth et al. 1972).

These observations give an indication of the physiological complexity accompanying circadian rhythms of activity and rest. But not every desert reptile shows day/night periodicity. The subterranean amphisbaenian *Diplometopon zarudnyi* of the Arabian peninsula shows no diel rhythmicity, even under experimental conditions of alternating light and darkness (Cloudsley-Thompson 1979b). In this respect, it differs from the nocturnal Indian sand boa *Eryx conicus*

Fig. 27. *Eryx conicus*, India

(Fig. 27) in which the light-dark cycle is an effective synchronizer of diel activity; more so than are fluctuations in temperature which affect the amplitude of the rhythm rather than the phasing of activity. Surprisingly, there is little evidence of the endogenous persistence of rhythm under constant conditions, but sand boas usually spend the daylight hours in an inactive state beneath the sand, although with their heads exposed so that they are not isolated from the daily cycle of light and darkness (Griffiths 1984).

3.4 Seasonal Rhythms in Locomotory Activity

Seasonal rhythms operate on two levels in regards to the avoidance of thermal extremes. First, they may influence the time of day at which locomotory activity engendered by the circadian rhythm of an animal takes place. Secondly, they often control the timing of aestivation, diapause, reproduction and other annual events (Sects. 6.4, 6.5). Many day-active desert arthropods and reptiles show bimodal rhythms of locomotory activity during the hot summer months, and a single midday peak at cooler times of the year.

3.4.1 Arthropods

In a study of the microhabitat (Sect. 3.2) of the desert woodlouse *Hemilepistus reaumuri* near Biskra, Algeria, little difference between temperatures in the air and on the ground was found in January. Temperatures ranged daily between 17 and 5 °C and no above-ground activity of *H. reaumuri* was observed. In March, however, temperatures on the ground surface rose to 32 °C during the day and woodlice were seen on the surface only in daylight with a peak of activity around noon. In July, surface temperatures reached nearly 70 °C while the air temperature 200 cm above the ground was 45 °C. Woodlice were then seen on the surface after dawn and at dusk. At a depth of 40 cm down their burrows the soil moisture never dropped below about 10% and the air within remained saturated at all times (Coenen-Stass 1989).

Although most millipedes are nocturnal, *Orthoporus ornatus* (Spirostreptidae) in the North American desert shows most activity during the early morning and late afternoon: some nocturnal feeding and locomotion also occur. Activity ceases before surface temperatures reach 35 °C, however, and begins again when they have cooled to about the same level. In this manner, heat and drought are largely avoided, the animals spending the time in the shade of vegetation or in cool, humid burrows (Wooten et al. 1975). Seasonal peaks in foraging have been recorded in subterranean termites, and are correlated with soil moisture content (Abushama and Al-Houty 1988).

Bimodal activity during hot weather has already been mentioned in the case of harvester ants (Sect. 2.6.1). It occurs in *Iridomyrmex purpureus* in Australia (Greenaway 1981), and Marsh (1987b) recorded the phenomenon among seven species of ants in the Namib desert. Colonies of *Pogonomyrmex barbatus*, however, show temporally patterned activities within the morning period in Arizona. Both the peak of nest maintenance and the onset of the peak of patrolling occur earlier than the peak of foraging (Gordon 1983). Of ten species of ants from the pre-Saharan steppe of Tunisia, three are day-active, two are nocturnal, while the rest show varying degrees of flexibility: they are active either by day or by night, but with shifts towards nocturnality in summer and diurnality in winter (Heatwole and Muir 1989). Larvae of tiger beetles, *Cicindela marutha*, in Arizona, are active at the surface during spring and autumn, but plug their burrows during the summer day and then become active only at night (Hadley et al. 1990).

Buxton (1923, 1924) commented upon the diurnal and seasonal appearance of many desert beetles, and showed that some large Tenebrionidae, such as *Adesmia* spp., which are very conspicuous in North Africa and Palestine at all times of the day in March and April, seek shelter under stones and bushes during the hotter months of May and June. Similar results have been obtained more recently by many other authors. Bodenheimer (1934) observed that the time of maximum activity of desert beetles varied with season. More recently, Alicata et al. (1979) described a reduction of locomotory activity in summer at midday in *Pimelia grossa* in the dune ecosystem of Isola delle Correnti, Sicily; Crovetti (1970) observed the same phenomenon in *Pimelia* spp. and *Blaps propheta* in Wadi Caam, Tripolitania; while Fiori (1956) and Mellini (1976a,b) recorded it among several other species of Tenebrionidae from the same region.

Similar observations have been made in the Namib desert, where the activity of *Cardiosis* spp. becomes bimodal on hot or calm days (Hamilton 1971). Timed traps and visual observations were used by Holm and Edney (1973) to measure the distribution of activity among other desert insects in southwest Africa. Not only did different species exhibit different seasonal and daily cycles, but summer and winter faunas were almost entirely different. *Onymacris rugatipennis* (Tenebrionidae) is bimodal in summer, but unimodal in winter. So, too, is *O. laeviceps* according to Hauffe et al. (1988). In the North American desert, likewise, modification of daily and seasonal activity in *Eleodes obscura* (Tenebrionidae) has been demonstrated by Marino (1986) who cites further examples. Apart from *Adesmia* and *Akis* spp., which are primarily day-active or crepuscular, most large Tenebrionidae of the Great Palaearctic desert, such as *Blaps, Pimelia* and *Trachyderma* (Fig. 19) spp., are nocturnal.

When the diel activity patterns of two subspecies of *Thermophilum sexmaculatum* (Carabidae; Fig. 20) from northern and southern regions of the Sahara were studied in the laboratory by means of aktograph experiments, it was found that the northern subspecies *T.s. sexmaculatum* adapts to changing seasonal climatic conditions by shifting its main activity phase within the 24-h period. This plasticity seems to have enabled the border regions of the northern Sahara to have been colonized from the south where *T.s. marginatum* is strictly nocturnal throughout the year, for beetles of the northern subspecies are diurnal when night time activity is rendered impossible by low temperature. In summer, however, they become nocturnal, thus avoiding the heat and dryness (Erberling and Paarmann 1985).

3.4.2 Reptiles

Similar seasonal shifts in the times of maximum activity have been recorded in the activity rhythms of many desert reptiles (Tinkle 1967; Mayhew 1968). They presumably facilitate thermoregulation by allowing animals to encounter similar thermal environments at different times of the year (Pianka 1977). In many cases, extreme daytime heat is avoided but, just as in the case of *Thermophilum sexmaculatum*, the winter cold at night may present a greater problem than the midday summer heat. For example, in southern Texas, desert tortoises, *Gopherus berlandieri* (Testudinidae), have two activity periods, morning and late afternoon in summer: they are inactive around midday. They are also inactive from December to February inclusive (Rose and Judd 1975). Likewise, the African *Geochelone sulcata* (Fig. 12b) is mainly active early and late in the day. It digs deep burrows for shelter during the heat of the day, and aestivates in them during the hot season (Grubb 1971). *Testudo kleinmanni* in Israel, is also active in winter but in summer retreats into rodent burrows and activity decreases (Green and Mendelssohn 1989).

The daily activity of the sand goanna *Varanus gouldii* (Varanidae) varies with the season, being greatest in summer and least in winter. Whereas the melanic subspecies *V. gouldii rosenbergi* of Western and South Australia is active throughout the year, however, *V. gouldii*, an inhabitant of semi-arid areas, remains underground during the winter (Green and King 1978). Seasonal changes in the diet and reproductive activity of *V.g. rosenbergi* have also been described by King and Green (1979). Fewer species of reptiles are eaten by this subspecies than by the desert subspecies *V. gouldii flavirufus* studied by Pianka (1970a). This probably reflects the greater number of reptile species which occur in the desert regions of Australia than in the coastal areas of South Australia, and suggests that both subspecies are general predators and scavengers — a characteristic of most desert carnivores.

Varanus gouldii is active throughout the year on Kangaroo Island, South Australia, whereas in arid and semi-arid regions it is inactive and remains in its burrow during the winter. It shows seasonal differences in the length of time spent basking before becoming active, in its heating rates and in its mean activity temperatures; but cloacal temperatures are maintained around 35.5 °C everywhere. The length of time during which the animal is active is dependent upon the ambient temperature and the time at which it emerges from its burrow. In arid regions, goannas are active throughout the day in early summer but, during hot weather, activity is usually restricted to early morning and late afternoon (King 1980). Although the American desert sidewinder *Crotalus cerastes* (Figs. 10a, 11) is predominently nocturnal, the speckled rattlesnake *C. mitchelli* is mainly diurnal in the spring and autumn and night-active during the summer months (Moore 1978). For further examples, see Section 6.4.2.

3.5 Social Behaviour

Social behaviour in desert Arthropoda has been reviewed by Cloudsley-Thompson (1989b). Evasion of both climatic extremes and of predators by burrowing is particularly important in the case of woodlice or sowbugs (Isopoda: Oniscidea; Sect. 3.2). For survival and reproduction, the desert woodlouse *Hemilepistus reaumuri* (Fig. 28a) of North Africa and the Middle East is dependent upon its burrows (Fig. 28b). These are vital to the animals' existence and have to be defended continuously against intraspecific and, to a lesser extent, interspecific competitors. Moreover, new burrows can only be dug during spring when the soil is moist. Such requirements are met by the cooperation of individuals within the framework of a complex system of social behaviour. Monogamous pairs of adult woodlice are formed in spring. The partners recognize each other individually through a highly developed system of chemical communication and later, with their progeny, form strictly closed family communities (Shachak 1980; Linsenmair 1984).

Although its physiology is remarkably well adapted to desert life, *H. reaumuri* would be unlikely to survive for a single day without the protection of its burrow from which it makes extensive foraging excursions. It usually returns from these by the shortest route, navigating during the day by the sun and the pattern of polarized light in the sky. The mechanism of orientation by external cues is backed up by idiothetic mechanisms by means of which *H. reaumuri* is able to forage even on moonless nights. These homing mechanisms involve a degree of error for which the woodlice compensate by systematic searching

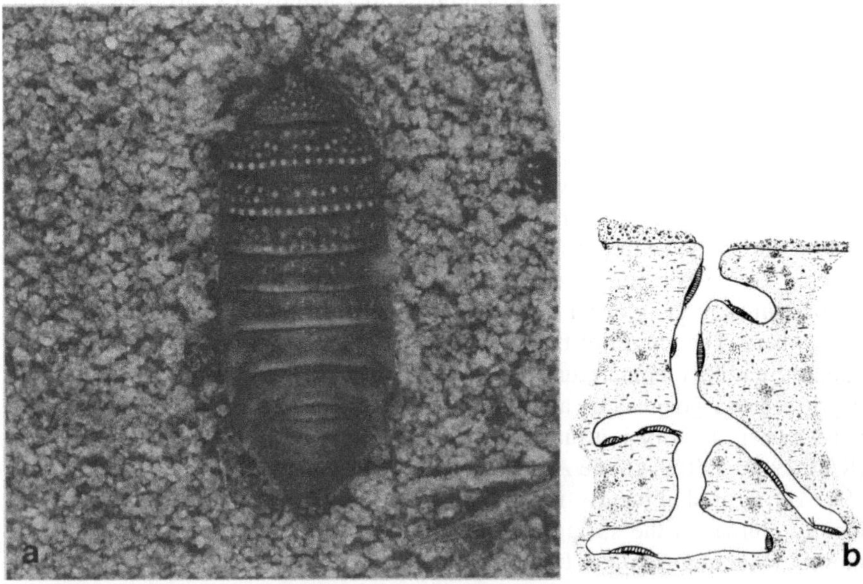

Fig. 28. **a** *Hemilepistus reaumuri*, Tunisia. **b** *H. reaumuri*; section of burrow (After Shachak 1980)

behaviour, locating the entrances to their burrows with the antennae (Hoffmann 1984).

Field observations carried out on four pairs of 11-month-old *H. reaumuri* in the Negev desert disclosed the fact that, during the gestation period, the amount of time spent above ground was short (male = 23 min, female = 17 min). The isopods allocated most of their surface activities to foraging and feeding: only a small fraction of the time was spent carrying food from the soil surface into the burrows. During the period of brood care, however, males and females increased their above-ground activity 5.2 and 7.6 times respectively, usually by carrying soil crust and plants into the burrows to feed their young (Shachak and Newton 1985). Undoubtedly, parental investment in brood care minimizes mortality in the juvenile woodlice and increases survival.

Few species of non-social desert insects construct their own burrows. Most secrete themselves beneath rocks and stones, while some large Tenebrionidae shelter in the burrows of scorpions. *Blaps sulcata* in Israel aggregates in large numbers under stones or crevices in the ground, mutually attracted by the exudations of their anal glands (Kaufmann 1966). Both scorpions and Solifugae, as well as lycosid spiders, are subsocial and protect their young until after completion of the first instar (Figs. 29, 30). This habit may be advantageous to desert living, but is not specifically an adaptation to arid environments since it is characteristic of all members of these orders, including the minority that inhabit more humid regions. Surprisingly, water appears to be exchanged between mother and first instar larval scorpions (Vannini et al. 1985). The mechanism by which

Fig. 29. *Leiurus quinquestriatus* with first instar young, Sudan

Fig. 30. Solifugid, *Galeodes granti*, guarding first instar young, Sudan

this occurs is not understood. Sociality among scorpions is discussed by Polis (1990b; See also Polis and Sissom 1990).

Termites and ants are successful insects of desert regions, to which they are pre-adapted because they are social and construct underground nests; although this, again, is not primarily a desert adaptation. Some termites may also construct mounds in arid regions, but the species that live in the hottest deserts tend to inhabit subterranean nests (Wallwork 1982).

Species of *Amitermes* in the Northern Territory of Australia build mounds that are orientated in a north-south direction (Fig. 31). The faces of the mound are thus presented towards the sun in the early morning and evening. This warms the nest at times when the air temperature is low whilst, at midday, a relatively small area of termite mound faces the sun. By cutting through such mounds at the base and rotating them 90 °C Grigg (1973) demonstrated that the orientation prevents overheating and results in a long midday temperature plateau. In winter, the termites aggregate on the east side in the morning and on the west in the afternoon; in summer, they retreat to the base or beneath ground level at midday.

It has recently been found that colonies fail after exposure to experimental alteration of the earth's magnetic field by means of bar magnets buried at the base of new nests. Whether termites use magnetic perception as a cue to achieving their spectacular compass orientation remains unknown, but the results of this field experiment weighs more for than against the possibility (Grigg et al. 1988). At the same time, if *Amitermes* spp. are indeed able to sense magnetic fields, the problem arises as to why they did not continue to build their nests in line with the artificial magnetic environment. Perhaps magnetic cues conflicted with others such as the daily movement of the sun, if this is perceived, and the colonies were handicapped by the receipt of conflicting orientation messages. Undoubtedly,

Fig. 31. Termite compass nests, *Amitermes* sp., Northern Territory, Australia

however, it is important to desert termites to be able to orient their mounds without ever observing the sun: compass orientation of termite mounds is not necessary in less extreme environments.

Colonies of harvester ants, *Pogonomyrmex barbatus*, also adjust their behaviour in response to changes in the environment. Ants flourish in deserts because large numbers of individuals comprising a single colony can excavate nests deep enough into the soil to reach relatively constant temperature and permanent water (Pierre 1958; Rissing and Pollock 1989). Workers change their tasks, or modulate the intensity of their activity, while the sizes of groups engaged in various tasks, and the amounts of time spent on them, are adjusted in response to the needs of the colony. Older colonies are less adaptable than younger colonies, and less able to adjust to worsening conditions (Gordon 1987). This may account for the disappearance of older colonies of ants and their replacement by younger colonies at times of stress, for adaptability is especially necessary for survival in desert regions.

Stegodyphus mimosarum (Eresidae) inhabits dry thornbush country in southern Africa. It is one of the most social of spider species. The nests are large, dense and compact. Young spiders hatched in a colony usually remain there. Nests are occupied and enlarged by consecutive generations of spiders, and may finally obtain the size of a man's head, acting as protective shields against predators, solar radiation and presumably also against excessive water loss in an arid environment. The size of the females at maturity is reduced in large colonies, as is the number of eggs produced per female. This reduction in fecundity apparently results from increasing competition for food and can be interpreted in terms of a 'constraint' model for group living. The high risk associated with departure tips the cost-benefit balance against reproduction (Seibt and Wickler 1988).

Most reptiles are not especially social, but interspecific (and intraspecific) responses to predators are probably quite widespread, especially among iguanid lizards, although there is, as yet, no evidence that they are especially marked in desert species. Parental care, too, is found throughout the class Reptilia (Shine 1988) but it is not an adaptation to heat, drought or exposure.

3.6 Conclusions

Most desert arthropods and many desert reptiles are nocturnal, spending the day in burrows and other sheltered retreats where the extreme daytime heat and drought are ameliorated. In this way they avoid the exigencies of the desert climate. Their times of activity are usually regulated by biological 'clocks'. Seasonal rhythms occur in the diel activity of some species, which are nocturnal in summer but are crepuscular or even day-active during the cool season of the year. Evasion of climatic extremes and predators can also be assisted by social behaviour.

4 Thermal Regulation and Control

4.1 Introduction

It is important to distinguish between thermal regulation and temperature control, as Bartholomew (1982) pointed out. Thermoregulation involves the maintenance of body temperature within a specified and usually fairly narrow range, whereas temperature control is much less precise. In the latter, rates of change may be modified, or a difference between body temperature and ambient temperature is maintained. In arthropods and reptiles, thermal regulation is achieved mainly by behavioural means: at the same time, there is often some capacity for physiological control of body temperature. The evolution of thermoregulation has been discussed by Whittow (1973).

Thermoregulation in arthropods has been reviewed, among others by Casey (1988), Cloudsley-Thompson (1964a, 1970b, 1975), Crawford (1981), Edney (1974), May (1979, 1985); in reptiles by Avery (1982), Bartholomew (1982), Bradshaw (1986), Cloudsley-Thompson (1971, 1972a), Dawson (1975), Heatwole (1976), Huey (1982), Mayhew (1968), Schmidt-Nielsen (1964), Schmidt-Nielsen and Dawson (1964), Templeton (1970), energetics and metabolism by Bennett (1982), Bennett and Dawson (1976). In the following, no attempt will be made to duplicate these comprehensive publications. Attention will be concentrated upon studies involving desert-inhabiting species, with special reference to recent work. The terminology of Bligh and Johnson (1973) has been adopted where applicable.

4.2 Behavioural Thermoregulation

Behavioural thermoregulation involves heat exchange with the environment by conduction, convection, radiation or evaporation. The first three of these exploit the differential between the surface temperature of an organism and that of its surroundings. They are influenced by the surface area, orientation and geometry of the body, its texture and, in the case of radiation, its colour; and by the conductivity of the environment, the wind speed, radiant intensity and so on (May 1979). The core body temperature (Tb) of a bradymetabolic animal therefore only reflects to a limited extent the air temperature (Ta), because it is also affected by longwave radiation, solar radiation and convection. Heat gain by conduction is almost negligible in arthropods and reptiles during the day, when it is restricted mainly to the ventral surfaces of the tarsi or feet. After dark, however, when the air cools, it may be the major source of heat gain, if not for arthropods then for nocturnal reptiles which press their bodies against the substrate.

The principal method by which ground-living desert arthropods and reptiles maintain their preferred body temperatures' or 'eccritic temperatures' is by moving or 'shuttling' between sources of warmth and 'heat sinks' where they shed surplus heat. The eccritic temperature is surprisingly close to the high lethal temperature. The significance of this is uncertain, although Heinrich (1977) has postulated that high temperature set points may have evolved from the inability to dissipate all the heat produced as a by-product of high rates of activity: many insects and some reptiles heat up considerably during strenuous activity. He concluded that biochemical restructuring for activity at high temperatures has evolved, in part, because it has extended the ability to be maximally active beyond the short time otherwise required to overheat. The higher the temperature set point, the greater the endurance of the high-rate aerobic activity at high ambient temperatures. Bradymetabolic animals exhibit adaptations for activity over a wide range of cell temperatures. At the same time, as Whittow (1973) pointed out, body size is at least as important in determining the character of thermal regulation as is taxonomy.

4.2.1 Insect Flight

Among the minority of day-active insects that spend their time directly on the hot surface of the desert, are a few species of wingless tenebrionid beetles, and 'velvet ants' (Multillidae), the females of which are also wingless. Presumably they are protected from insolation by their dense covering of hair. After the sand temperature reaches about 50 °C, however, they seek shelter in the shade of vegetation. Some beetles and lizards maintain themselves in a tolerable environment by moving vertically up and down the stems of desert plants as the air temperature increases or decreases.

Flying insects are unique in carrying vertical migration to an extreme. Desert grasshoppers and locusts may at first extend their long legs and raise their bodies into a perceptibly cooler stream of air. As the temperature increases, however, they will fly a few metres and are further cooled whilst aloft. Predatory tiger beetles (Cicindelidae), beeflies (Bombyliidae; Fig. 71), and robber flies (Asilidae) alight only briefly on the hot sand, and then cool off as they range above its surface in search of food.

Of all the insects that spend part of the time on the surface of the desert soil and then cool off whilst in flight, possibly none are more striking than digger wasps (Sphecidae). These insects face a special problem because, since they hunt other insects by sight, they have to dig their burrows and provision them during the hottest hours of the day. They are not insulated with hair as are Mutillidae and so are readily susceptible to overheating; they rapidly succumb to heat stupor if restrained on the hot soil surface. But they hover in the air, descend almost vertically to the site of the burrows and dig briefly with their curved front feet. Then they rise again into the air, hover briefly, and drop down again to resume their excavations (Amos 1959). The effects of metabolic heating (Sect. 4.4.1) and evaporation cooling (Sect. 5.2.3.1) during flight will be considered later.

4.2.2 Moving and Shuttling Between Sunlight and Shade

The primary method of thermal regulation employed by all day-active desert poikilotherms consists of shuttling between sunshine and shade. The literature is replete with examples. Because terrestrial environments are thermally heterogeneous, small desert animals are able to select areas within their habitats that allow them to stabilize their Tb near optimal levels by moving between sunny and shaded substrates. In the sun, further control may be possible by altering their orientation (Sect. 4.2.3).

4.2.2.1 Arthropods

An excellent example of behavioural thermoregulation is provided by the flightless black desert grasshopper *Taeniopoda eques*, a native of the Chihuahuan desert. At dawn, grasshoppers descend to the ground from their nighttime roosts in bushes. They return to their roosts during the middle of the day, move back to the open ground in the afternoon, and re-ascend the vegetation at dusk.

Thermoregulatory postures are synchronized with these changes in microhabitat. During the cooler mornings and afternoons, they maximize heat gain by turning their flanks to the sun and crouching, thereby achieving thoracic temperatures up to 16 °C above the ambient, while at noon they stilt in the shade, minimizing heat gain. This behaviour keeps Tb near the preferred temperature (36.2 °C) but lower than the voluntarily tolerated temperature (41.9 °C), critical thermal maximum (CTM; 45.2 °C) and instantaneous lethal maximum (46.5 °C). Thermoregulation is influenced by size and colour (Sect. 4.4.2) which, to some extent, is the product of a reliance on chemical defence. In other words, for *T. eques*, feeding, oviposition, mating, anti-predator defence, size, colour and thermoregulation are all interrelated. Each in some way influences and, in turn, is influenced by the others, and all resonate to the physical constraints of the desert environment (Whitman 1987, 1988).

Many other desert arthropods have been studied with reference to the regulation of body temperature by moving between sunshine and shade, chief among them being tenebrionid beetles. In the Namib desert, for example, day-active beetles are mainly species of *Onymacris* (Fig. 32a,b), *Stenocara* and so on, although insects rarely shuttle rapidly between sun and shade as some lizards do (Edney 1971a). Such behaviour could stabilize Tb, but the frequency of sun-shade transitions would have to be extremely high, owing to the low thermal inertia of insects. Desert beetles, likewise, vary in their selection of sunlit sites. *Onymacris rugatipennis* (Fig. 32b) is active primarily in sunny areas in the early morning at

Fig. 32. a *Omymacris plana*, Namib desert. **b** *O. rugatipennis*, Namib desert

low Ta, and primarily in the shade at high Ta, as also is *O. plana* (Figs. 32a, 33), and *Eleodes* spp. in the North American desert (Kramm and Kramm 1972). As Ta increases, several species of day-active Namib desert tenebrionids move from directly insolated dune faces to shade, or to dune faces where the angle of insolation is lower (Hamilton 1971; Henwood 1975). Thus, in *O. plana* (Figs. 32a, 33) for instance, thermoregulation encompasses a number of behavioural reactions and postures. The peak of activity is shifted on hot days from midday to dawn and dusk, while excessive heating is avoided by burrowing. This species has long legs and is very speedy, sprinting between clumps of vegetation during the heat of the day when wind velocity is low and radiation high (Sect. 7.3.1). In

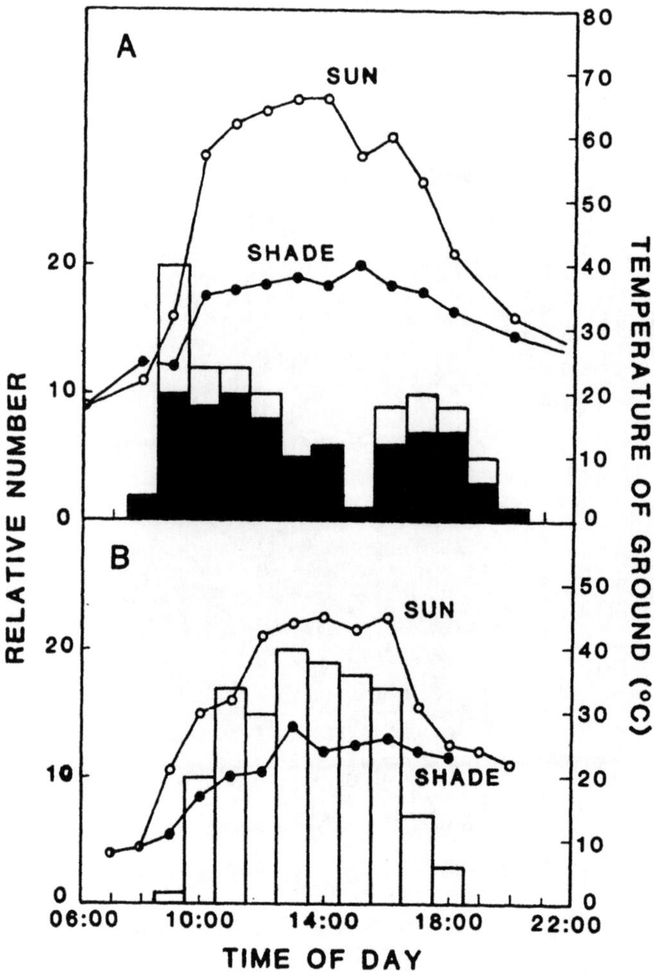

Fig. 33 A,B. Distribution of activity of *Onymacris plana* (*histograms*) and associated ground temperatures (*points*) during summer (**A**) and winter (**B**). In upper histogram, *filled bars* indicate beetles in shade, *open bars* beetles in sun; no distinction is made in lower histogram. Relative numbers are not comparable between seasons (May 1985, after Holm and Edney 1973)

this way, convectional cooling may be increased considerably. Less significant is posture. Nevertheless, the beetles crouch against the ground on cold mornings, pressing their bodies against the warm surface of the sand: later in the day they stilt (Sect. 4.2.3), thereby increasing convection.

Of the desert insects that control Tb by moving between sun and shade, locusts are among the best studied (Uvarov 1977). The desert locust *Schistocerca gregaria*, for example, avoids shaded areas at low Ta but aggregates in the shade at high Ta (Ellis and Ashall 1957), and strong negative correlations have been established by Anderson et al. (1979) between Ta and the relative numbers of *Psoloessa delicatula* and *Eritetrix simplex* in sunlight. At very high Ta, desert locusts may shelter in rock crevices (Uvarov 1977), while the grasshopper *Sphingonotus carinatus* orients itself lengthwise in the narrow shadows of plant stems (Hafez and Ibrahim 1964). Such movements may be a direct response to temperature, as in the case of locusts, or they may be thermally mediated responses towards distinct areas of shade, as in desert beetles. Many insects also change from positive to negative phototaxis as temperature increases (May 1985).

In addition to moving between shade and sun, ground and vegetation, one part of the burrow and another, sand-dwelling arthropods move nearer to, or further from, the surface according to the gradients of temperature and humidity that they experience. A fine example is afforded by the well-known North American desert cockroaches *Arenivaga investigata* (Dictyoptera). These insects live in sand dunes with less than 1% moisture most of the year. During the day in spring, summer and autumn, they burrow at a depth of 20-60 cm while, at night when surface temperatures have cooled, they come to within 1-3 cm of the surface (Sect. 3.2). In winter they remain active at lower levels during both day and night (Hawke and Farley 1973; Edney et al. 1974).

The original field experiments were subsequently confirmed experimentally by Edney et al. (1978), who set up horizontal and vertical gradients of temperature and humidity in sand columns, and monitored the conditions within them by appropriate sensors. *Arenivaga* nymphs, tagged with a radioactive isotope, were introduced, and their movements observed and photographed by means of a gamma-scintillation camera. In this way, it was possible to determine that the insects oriented themselves in both horizontal and vertical gradients of temperature and humidity, showing a preference for a higher moisture content and for temperatures in the range 22-36 °C, with maximum aggregation from 27-29 °C. The response is orthokinetic, and the sensillae involved are probably situated on the antennae.

The most important strategies for survival of trogid carrion beetles (*Omorgus, Trox* and *Polynoncus* spp.) in the arid Kalahari include a biphasic daily pattern of activity which coincides with peaks in relative humidity and the inverse in temperature, the ability of adults and larvae to enter a quiescent state under adverse environmental conditions, rapid immature development, long adult life and consequent overlapping of generations, suspended ovariole development and a tendency to a flightless condition (Scholtz and Caveney 1988). Once again, it can be seen that adaptation to the desert environment is not restricted to a limited number of parameters but affects every aspect of behaviour, physiology, morphology and life history.

Burying behaviour, however, is not invariably associated with the need to attain preferred body temperature or to avoid desiccation. Thus, tenebrionid beetles inhabiting the dunes of the Namib desert spend most of their lives buried below the surface of the sand. When they do surface, which is not necessarily daily and is often for only a few minutes or hours, the temperatures available are such that, had they remained buried, they could have attained their preferred body temperatures of 36-40 °C and simultaneously avoided solar radiation. At the same time, desiccation stress is greater on the surface during the day at the time of emergence, and greater below the surface at night when they are buried (Seely and Mitchell 1987).

4.2.2.2 Reptiles

Early studies on thermoregulation in reptiles (Mosauer 1936; Cowles and Bogert 1944; etc.) produced convincing evidence that most species maintain preferred body temperatures by behavioural means. The vast literature on the subject, cited in the various reviews mentioned above (Sect. 4.1), clearly documents the complexity of the phenomenon. In cool weather, heliothermic reptiles may often be compelled to operate at temperatures below the optimum while, in hot deserts, the environmental temperature may be so great during the day that the animals are unable to prevent their Tb from rising above preferred levels. This has been amply demonstrated, for instance, in the Australian lizard *Amphibolurus inermis* by Heatwole (1970, 1977).

The CTM is the temperature at which locomotory disability develops. It is probably one of the best indirect correlates of optimum Tb during activity (Huey 1982) but it does not take the breadth of thermal performance into account. Several indirect estimates of this have been proposed, but none have yet proved to be consistently reliable. Dreisig (1985), however, has recently presented a mathematical time-budget model for thermoregulatory basking in shuttling ectotherms, based upon physical laws of heating and cooling, and on the assumption that upper and lower limits for Tb exist. This model is supported by observations of basking in a tiger beetle (Cicindelidae) and the European lizard *Lacerta agilis*.

Land tortoises bask in the sun until their deep body temperatures have reached the preferred level, and may voluntarily permit them to approach the CTM. Only then do they leave their basking sites and retire into the shade. This subject has been reviewed by Boyer (1965), Cloudsley-Thompson (1971), Sturbaum (1982) and others. Many semi-desert and savanna species, such as the North African *Geochelone sulcata* (Fig. 12b) dig deep tunnels for shelter and aestivation (Grubb 1971). So, too, does *Gopherus agassizii* (Fig. 12a). According to Stebbins (1954), the thermal adaptations of *G. agassizii* to desert life also include the construction of these burrows and the thick, insulating shell. At the time of midday retreat to its burrow, the temperature of the surface of the carpace of this species may be more than 10 °C above deep body temperatures, thus demonstrating that the shell acts as an effective buffer against solar radiation (McGinnis and Voigt 1971).

A comparative study on behavioural and physiological thermoregulation in three species of southern African tortoise indicates that *Homopus areolatus*, the smallest, has greater thermoregulatory capacities through behavioural means than *Chersina angulata* or the large *Geochelone pardalis*. *H. areolatus* also possesses greater physiological control of conductance, and has a higher maximum preferred temperature than the other two species. In contrast, big specimens of *G. pardalis* show little physiological control of heat flux. Although the three species occur sympatrically, their thermoregulatory mechanisms differ; this cannot be explained in terms of size alone (Perrin and Campbell 1980).

More is known of the desert adaptations of lizards than of any other reptilian taxon. Thermal adjustments include elevated CTM and preferred temperatures which are reflected in behavioural differences between congeneric species and different populations of the same species (Mayhew 1968; Arnold 1984b). For instance, five North American desert species of Iguanidae: *Callisaurus draconoides*, *Gambelia wislizenii*, *Dipsosaurus dorsalis*, *Sauromalus obesus* and *Uma notata*, maintain very high Tb (above 38 °C) for extended periods. Of these, *D. dorsalis*, in particular, frequently tolerates temperatures close to the CTM and well above those of species from cooler environments (Dawson 1975).

Closely related reptile species that have a relatively similar mean Tb may be utilizing temperature in a slightly different manner. For instance, *Uma notata* appears to tolerate a wider range of Tb while active than do either *U. inornata* or *U. scoparia*, and seems to prefer a slightly higher temperature even though the mean Tb is approximately the same for all three species (Mayhew 1964). On the other hand, different sympatric desert lizards apparently have distinct, predictable thermoregulatory behaviour patterns which account for the differences in their Tb (McGinnis and Falkenstein 1971).

Greer (1980) found that lizard species from the warm and interior regions of Australia had a CTM above 39.5 °C, irrespective of their behaviour or habitat, while species from the cooler and coastal regions could be divided into two groups: diurnal surface-dwelling forms usually had a mean CTM above 39.5 °C, whereas crepuscular, nocturnal, cryptozoic and burrowing species had maxima below this temperature. Again, specimens of *Sceloporus graciosus* from higher altitudes (23 000 m) in Wyoming show preferred temperatures some 3 °C lower than lizards collected at altitudes of 1600 to 1650 m (Mueller 1970).

Berk and Heath (1975) used a shuttle box to determine the high and low body temperatures at which the North American desert iguana *Dipsosaurus dorsalis* shuttles between hot and cold environments. The average means were 41.7 ± 1.66 °C and 36.41 ± 1.12 °C respectively. These temperatures are within the range of Tb found in desert iguanas in a thermal gradient (McGinnis 1966; De Witt 1967) and are similar to the Tb of active lizards in the field (Cowles and Bogert 1944; Norris 1953).

The flexibility of thermoregulatory behaviour in the western fence lizard *Sceloporus occidentalis* is an adaptation that allows greater habitat utilization than would otherwise be the case (McGinnis 1970). The related *S. magister* is usually arboreal and found in association with a wide variety of trees and larger cacti, but is sometimes terrestial or saxicolous. Body temperatures are maintained over a fairly narrow range (30.3 - 37.8 °C) and average 34.8 °C (Parker and Pianka 1973). In a recent investigation of the thermal biology of the related *Sceloporus merriami* in Texas, Grant and Dunham (1988) found that the thermal environment had a major impact on this species whose Tb (on average 32.2 °C) is lower than those of other North American desert iguanids. The maximum of activity, around sunrise and sunset, coincides with the time when the average environmental temperature approximates the mean body temperature of the lizards. Also, at this time, the lizards occupy thermal microclimates randomly on their home ranges whereas, when activity rates are lower, their use of microclimates is not random. Thus, when thermally favourable microclimates are most available, activity is least constrained.

Skinks, once thought to be lizards of cooler environments, are now known to include many high temperature desert species (Greer 1980). Pianka (1969b) found that whereas Western Australian desert skinks, *Ctenotus* spp., which forage in the open, show a poor correlation between Ta and Tb, those that forage in shady places within vegetation have a Tb which is closely correlated with Ta. This is because the former depend upon direct solar and substrate heat, while the latter do not appear to exploit these heat sources substantially.

Thermal responses, oxygen consumption and evaporative water loss are related to both habitat and burrowing behaviour in scincid lizards. They are lowest in desert species such as *Typhlocontias brevipes*, a legless, nocturnal, burrowing skink of the Namib desert dunes (Withers 1981).

Brattstrom (1971), too, found a marked difference in the thermal patterns and tolerances of Australian skinks, reflecting their radiation into cool-moist, warm-moist and hot-arid environments. Even after acclimation, the CTM was highest in the desert species. Schall (1977) showed that southwestern Texas teiids (*Cnemidophorus* spp.) display an evolutionary conservative thermal tolerance, despite radiation into a variety of habitat niches.

In a comprehensive study of the thermal biology of chameleons, Burrage (1973) made a comparison between the arboreal *Chameleo pumilus* and the terrestrial desert-dwelling *C. namaquensis* (Fig. 34) in southern Africa. He found that *C. pumilus* was active under a wide range of weather conditions, and would even hunt for food while it was raining, whereas *C. namaquensis* was a sun basker, with less variable body temperatures and more restricted times of activity. Its

Fig. 34. *Chameleo namaquensis*, Namib desert

range of activity varied between 14 and 39.7 °C (mean 28.7 °C). The range of activity of *C. pumilus* could be as low as 3.5 °C or as high as 37 °C (mean 22.4 °C). These envorinmental considerations appear to be mainly responsible for viviparity in *C. pumilus* and oviparity in *C. namaquensis* which lays its eggs some 13 cm below the desert substrate where temperature and moisture conditions are remarkably stable.

Experimental sequences, in which animals were deliberately heated and cooled in a gradient, revealed considerable thermoregulatory capability in the legless Australian pygopodid *Lialis burtonis*, whose preferred Tb of 35.1 ± 0.11 °C did not differ over an appreciable time span (Bradshaw et al. 1980). The limited thermoregulatory choices available to nocturnal geckos, on the other hand, result in low mean activity temperatures, a wide activity range and a great deal of variation around the mean. The term 'preferred temperature' is probably not applicable in discussing the ecology of these eurythermic reptiles according to Marcellini (1976).

Most studies of the thermal ecology of desert lizards have been conducted on day-active species, although Pianka and Pianka (1976) investigated air and body temperatures, and discussed the significance of the preferred temperatures of 12 species of nocturnal geckos in the Western Australian desert. In the case of *Coleonyx* spp. (Gekkonidae) of Mexico and Texas, Dial (1978a,b) found than *C. brevis* is active at night at suboptimal temperatures and has adopted a eurythermic strategy, thereby exploiting successfully its nocturnal behaviour. In contrast, *C. reticulatus* is active at considerably lower temperatures and inhabits comparatively cool, sheltered crevices. *C. brevis* thermoregulates behaviourally while under cover, pressing its back against the lower surfaces of rocks warmed by the sun.

The mean eccritic temperatures of *Acanthodactylus boskianus* and *A. scutellatus* in Israel have been found to be the same. These species have similar body lengths. Significant differences occur, however, between *A. schreiberi* and the smaller *A. pardalis*, which also has a higher rate of oxygen consumption (Duvdevani and Borut 1974). The legless fossorial lizard *Anniella pulchra*, which inhabits the Pajarro dunes of California, shows a thermal preference for 24-25 °C avoiding temperatures below 20 and above 28 °C. It apparently has a broad thermal range and prefers higher temperatures under moist conditions. A comparatively low thermal preference may permit increased activity in its subterranean environment where ambient temperatures are often cool (Bury and Balgooyen 1976).

Tropical and subtropical deserts may be extremely hot by day, but they can also be surprisingly cold at night. Thus, Porter and Gates (1969) have calculated that, at night and in still air, *Dipsosaurus dorsalis* could not survive a temperature below 10 °C if exposed to a clear sky, and not less than 5 °C at a wind speed of 100 cm s^{-1}. A burrow would have to be 3 °C warmer for the lizard to survive. In full sun, in still air, it could not survive a Ta below -12 °C if dark in colour or -7 °C if pale. At a wind speed of 100 cm s^{-1}, in sunshine, it could not withstand less than -2 °C if dark or 0 °C if light in colour (see also Sect. 3.4.2). This, however, is a wind speed higher than is normally encountered near the ground.

Most reptiles, including desert species, tend to show an increase in their eccritic temperatures after feeding. This probably facilitates digestion; it has been demonstrated in the North American box turtle *Terrapene ornata* by Gatten (1974). The skink *Lygosoma laterale* congregates in warm areas when it has eaten, while thermophilic responses after feeding have been noted in North American alligator lizards, *Gerrhonotus multicarinatus*, and the night lizards *Xantusia vigilis* and *X. reversiana* by Regal (1966); and in *Amphibolurus nobbi* by Witten and Heatwole (1978). According to Bustard (1968b), metabolic activity and digestion in the Australian gecko *Gehyra variegata* are greatly depressed by cool night temperatures. Mackay (1968) studied the effect of temperature on fluctuations of the gastrointestinal pressures of *Varanus flavescens* and *Ctenosaura pectinata*. He found that the frequency of contractions decreased at lower temperatures, below those at which feeding takes place. The feeding of *Aporosaura anchietae* of the Namib is also closely related to its thermoregulatory behaviour (Louw and Holm 1972) as, probably, similarly to that of other desert lizards.

The state of nutrition is not the only physiological factor influencing thermal responses in lizards. For several hours after exercise, *Dipsosaurus dorsalis* selects warmer ambient temperatures than normal, resulting in small but significant increases in Tb (Cannon and Kluger 1985). Conversely, thermal optima may reflect other physiological functions. Thus, Werner (1983) found a good overall correlation among lizards between specific optimal temperatures for cochlea performance and ecological, whole body, optimal and preferred body temperatures. This is part of the well-known phenomenon in which many physiological processes in reptiles tend to have their optima at or near the ecological preferred Tb (Huey 1982; Werner 1985). Furthermore, there may be a relationship between temperature and morphological characters. Regal (1975) reported a correlation in several genera of lizards between environmental temperature and the sizes of the scales. Small scales appear to be selectively advantageous in cool environments, large scales where heat is a problem.

In general, desert snakes are unable to withstand body temperatures as high as those tolerated by lizards, and their CTM is considerably lower (Mayhew 1968; Cloudsley-Thompson 1971; etc.; Fig. 35). Although the Californian striped racer *Masticophis lateralis* shows a pattern of thermoregulatory behaviour similar to that of various heliothermic desert lizards, it emerges with a low Tb, basks before its morning activity, avoids the midday heat and retreats in the afternoon with a high Tb. No significant differences have been found in Tb between fasting and recently fed snakes (Hammerson 1979). The bull snake *Pituophis melanoleucus* (Fig. 36) does the same in less rigorous environments although it tends to be crepuscular or even nocturnal in deserts (Mayhew 1968).

4.2.3 Posture and Orientation

Although fossorial habits, coupled with nocturnal activity (Sect. 3.3) and shuttling between sunlight and shade (Sect. 4.2.2), are the principal means by which

Fig. 35. Western diamond-back rattlesnake, *Crotalus atrox*, sheltering from the sun, Arizona

Fig. 36. Bull snake, *Pituophis melanoleucus* (Colubridae), seeking shelter after basking, New Mexico

bradymetabolic desert animals maintain thermal homeostasis, posture and orientation also play an important part in behavioural thermoregulation. The classical example is afforded by locusts which orient their bodies at right angles to the sun's rays at low Ta during the morning and evening, but face the sun at the hottest time of the day, thereby reducing the area exposed to insolation. Moreover, by forming dense aggregations in sunny areas they effectively reduce their surface to volume ratio and increase the boundary layer of the group, thus allowing maintenance of a higher Tb than in solitary individuals (Chapman 1959).

Mention has already been made to *Taeniopoda eques* (Sect. 4.2.2.1) in which changes in microhabitat are supplemented by the adoption of thermoregulatory postures. These include orienting broadside to the sun, a posture sometimes known as 'flanking', 'stilting' with thorax and abdomen elevated above the substrate, 'crouching' with the venter of the body pressed against a warm substrate such as a rock and 'stem-shading'. Grasshoppers exhibiting this behaviour cling to the shady sides of stems with their venters directed towards, but shaded from, the sun (Whitman and Orsak 1985).

In their study of habitat selection in two species of grasshoppers, *Psoloessa delicatula* and *Eritetrix simplex*, in Colorado, Anderson et al. (1979) unravelled an elaborate repertoire of thermoregulatory postures and shade-seeking behaviour (Sect. 4.2.2.1). *P. delicatula*, which inhabits short grass prairie with bare patches, regulates Tb very effectively. In contrast, *E. simplex* inhabits denser vegetation where behavioural thermoregulation is not possible, loses water rapidly by transpiration, and Tb equals Ta.

Comparable postures are probably adopted at high temperature in many other desert arthropods. Stilting, or elevating the abdomen, has been described in a number of scorpion species, but Alexander and Ewer (1958) were the first to demonstrate that in the South African *Opisthophthalmus latimanus* stilting is exhibited during the hot hours of the day when the scorpion waits in the entrance of its burrow to catch prey. Whilst stilting, the legs are straightened, lifting the body clear of the substratum, as a result of which circulation of air around the animal is increased. When the temperature becomes so high that stilting no longer has any protective value, a photopositive reaction, which normally keeps the scorpion at the entrance of its burrow, changes to a photonegative one so that the animal retreats into the cool depths within.

Thermoregulatory behaviour of the vegetarian lizard *Angolosaurus skoogi* (Cordylidae) of the Namib has been described in detail by Hamilton and Coetzee (1969). It consists of subsurface basking, slip-face basking, crest basking and plunging deep into the sand. There is also a range of postural adjustments when the feet, body or tail may be raised above the hot sand. *Aporosaura anchietae* (Lacertidae) does much the same, as does the Australian *Amphibolurus ornatus*. These reactions are highly adapted to the unique environment. The general behaviour is similar to that of most day-active lizards living in vegetated regions in terms of basic requirements except that, for *A. skoogi*, cool subsands are substituted for the shade of vegetation or cool burrows. *A. anchietae* is a sand-diving lizard of the Namib desert, which inhabits the soft, wind-blown sand of the dune slip face. This lacertid is strictly day-active (Holm 1973) but, nevertheless, escapes extremely high and low surface temperatures by adopting a bimodal activity rhythm which coincides with surface sand temperatures of 30-40 °C (Louw 1972; Louw and Holm 1972).

More recently, Seely et al. (1988a) continued the investigation of Hamilton and Coetzee (1969), and related posture, locomotion and other aspects of the surface activity of *Angolosaurus skoogi* to the microclimate prevailing above and below the sand surface. As a result of the effects of radiation and wind, the subsurface sands and the slip face provide a thermal mosaic of which *A. skoogi* is able to take advantage. The animals modify their surface activity according to the prevailing microclimate, showing a trend towards a bimodal activity pattern on warmer days (Sect. 3.4.2). Thermoregulatory behaviour is facultative. Emergence is governed primarily by an innate circadian rhythm but, after this has played its part, activity can be modified extensively according to prevailing microclimatic conditions.

Greenberg (1977) devised an ethogram of the blue spiny iguanid lizard *Sceloporus cyanogenys*, incorporating an inventory of no less than 76 distinctive postures, orientations and simple movements. In it, thermoregulatory behaviour was divided into basking and thermal-comfort patterns. The theoretical arguments provided by Norris (1967) and Porter and Gates (1969) suggest that orientation to the sun has a greater influence on the Tb of larger than of smaller zebra-tailed lizards, *Callisaurus draconoides* (Iguanidae). Muth (1977a,b) concluded that thermoregulatory postures and orientation to the sun have important effects on steady-state body temperatures.

4.3 Thermal Tolerance

4.3.1 Heat Tolerance

Although the avoidance by behavioural means of temperatures that are lethal or cause irreversible damage is of prime importance to desert animals, even in their refuges and retreats they may experience wider extremes than do comparable forms in temperate regions or the humid tropics. To some extent this problem is countered by greater thermal tolerance. Thus, it is well known that many species of animals normally inhabiting warm environments are adapted to temperatures higher than those tolerated by the inhabitants of cooler regions (Andrewartha and Birch 1954). Seely et al. (1988b) found that the body temperatures of diurnal Namib dune tenebrionids, which are active on sandy substrates throughout the year, range between 30-39 °C. They pointed out that the ability to withstand high body temperatures is characteristic of desert arthropods (Cloudsley-Thompson 1970b; Edney 1971a). At the same time, two other factors may have led to their high preferred body temperatures: the aseasonality of the Namib environment and the presence of a readily available thermal refuge, i.e. the subsurface dune environment.

After a year of intensive field study, Collins et al. (1973) established foraging limits between 7.6 and 47 °C for the termite *Heterotermes aureus* and 9-49 °C for *Gnathamitermes perplexus* in the Sonoran desert, Arizona. Curtis (1985a) found by experiment that the mean preferred temperature of workers and brood of the Namib desert dune ant *Camponotus detritus* (Fig. 37) was 35 °C at 100% RH, 31-33 °C at 30% RH, while the CTM was as high as 53 °C at both relative humidities when the temperature was rising 1 °C min^{-1}. The ability to tolerate 45 °C for up to 24 h was not affected by humidity. The critical minimal temperature was 4.57 °C at 100% RH, and workers tolerated -1 °C at 93% RH and 45 °C at both 95 and 45% RH for 24 h. The sub-Saharan *Cataglyphis bombycina* can withstand 45 °C indefinitely and 50 °C for more than 10 min (Heatwole and Harrington 1989). Again, *Ocymyrmex barbiger* has a CTM of 51.4 while the sand surface temperature is 67 °C (Marsh 1985). The larva of the Namib desert ant-lion *Cueta trivirgata* also tolerates high body temperatures (LT_{50} = 53.4 °C). By exploiting the microclimate of its pit, and by digging below the surface when the heat is extreme, it can capture prey at surface temperatures of 13-63 °C (Marsh 1987a). The American desert wolf spider *Lycosa carolinensis* compensates for high summer temperatures and wide seasonal variations of daily temperature by rapid thermal acclimation, a relatively low Q_{10} below 39 °C, a lateral shift of the metabolic rate-temperature curve, and active selection of high relative humidities when the ambient temperature is high (Moeur and Eriksen 1972). Lethal temperatures (LT_{50}) for exposures of 24 h below 10% relative humidity as high as 50 °C have been recorded in the Sudanese solifugid *Galeodes granti*, 47 °C in the scorpion *Leiurus quinquestriatus*, 43 °C in the tenebrionid beetle *Trachyderma* (= *Ocnera*) *hispida*, and 43 °C in *Pimelia grandis* (Cloudsley-Thompson 1962c).

In the case of reptiles, Curry-Lindahl (1979) concluded that desert species and subspecies show greater thermal tolerance than do species living in equatorial

Fig. 37. Namib dune ant, *Camponotus detritus*

regions. The family Agamidae, in particular, appears to be best adapted to high Tb, with 13 species (of 18 tested) having lethal temperatures above 43 °C, in comparison with six species in the Iguanidae (of 33 tested) and four in the Gekkonidae (of nine tested). Agamids, iguanids, teeids and geckos, in hot environments of five continents, converge along the same lines behaviourally and physiologically, in regards to their responses to territoriality in combination with high temperatures, a feature that is of adaptive significance.

As Gans and Dawson (1976) pointed out, even where the maintenance of an elevated Tb carries with it a substantial benefit, the high eccritic temperatures of desert lizards may represent not so much a physiological need as a thermal level that minimizes dependence on the use of 'heat sinks'. Such minimization in itself, however, may require adjustments at the cellular level (Dawson 1975). The costs and benefits of thermoregulation in lizards have been evaluated by Huey and Slatkin (1976), who concluded that the physiologically optimum temperature is not always the ecologically optimum one. Thermoregulation is beneficial only when associated costs are low: thermal specialists normally thermoregulate more precisely when the productivity of the habitat is increased or when competition is reduced.

In a penetrating study of behavioural attitudes and thermal regulation in five species of *Amphibolurus* (Agamidae) in a variety of arid and semi-arid areas of Australia, Bradshaw and Main (1968) distinguished several different patterns of behaviour as being of significance in thermoregulation; these are summarized in Fig. 38. They concluded that selection for the ability to tolerate increased body temperatures does not appear to have operated in the genus *Amphibolurus*, due to the efficiency of the thermoregulatory behaviour of these lizards. Desert-living species show no greater tolerance of grossly elevated body temperatures than do species from less severe habitats when their Tb is similarly elevated, although the former are exposed to environmental temperatures much greater than those prevailing in semi-arid and subhumid habitats. All species depend upon thermal refuges for their survival during the hottest periods of the day, but especially those inhabiting very hot, dry regions. In contrast, *Diporophora bilineata*, another sympatric agamid, which is unable effectively to avoid the heat of the day, shows an ability to withstand body temperatures far in excess of those tolerated by most thermophilic species of *Amphibolurus*. These results contrast with those of Heath (1965) who found that desert species of *Phrynosoma* consistently showed thresholds for high-temperature behaviour patterns 2-3 °C above the thresholds for non-desert species.

Many day-active lizards maintain comparatively high body temperatures by means of thermal regulation as we have seen. They should theoretically, therefore, have higher daily integrated body temperatures, and hence higher daily energy and food requirements than nocturnal lizards which have lower body temperatures when active. This hypothesis was tested by Nagy and Knight (1989) with a diurnal skink, *Mabuya striata*, and a nocturnal gecko, *Pachydactylus bibroni*, both of which are arboreal insectivores, in the Kalahari desert. Doubly labelled water measurements (in which the lizards were given intraperitoneal injections of sterile water containing oxygen-18 and tritium) indicated that field metabolic

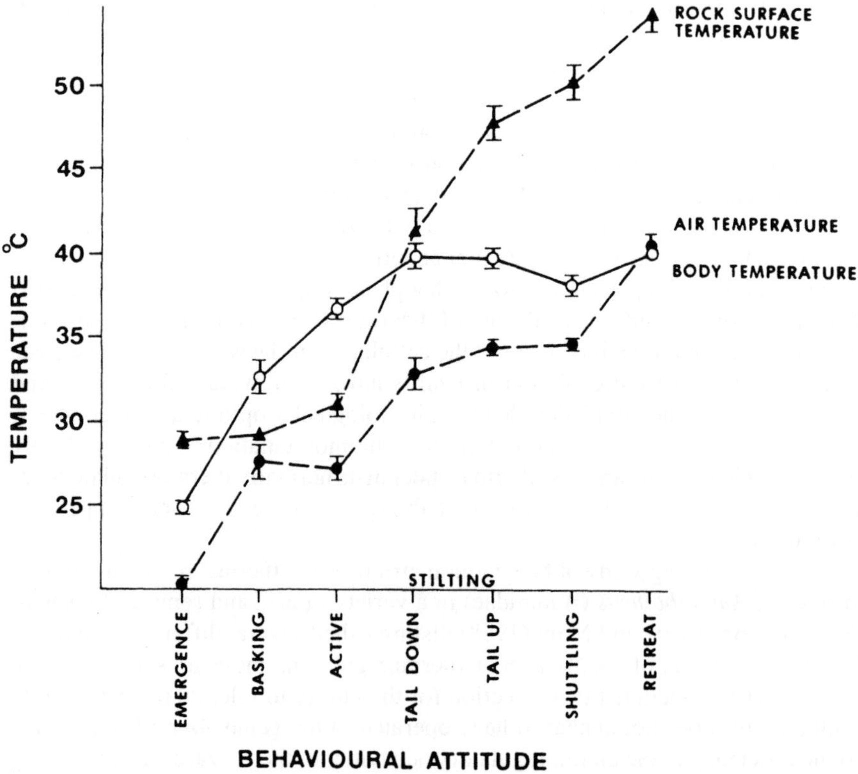

Fig. 38. Relationship between mean body temperature, associated with various thermoregulatory behaviour patterns of the lizard *Amphibolurus ornatus*, and associated environmental temperatures (plotted from data in Bradshaw and Main 1968) (Bradshaw 1986)

rates, water flux rates and feeding rates do not differ between these two sympatric species. The unexpectedly high metabolic rate of the geckos may, in theory, be due to a higher Tb during the daytime, greater activity whilst abroad, longer periods of activity or different thermal effects on energy metabolism than previously thought.

4.3.2 Maxithermy

Behavioural thermoregulation by day-active desert arthropods and reptiles permits the maintenance of elevated body temperatures for long periods. Since the metabolic rate of ectotherms rises with Tb, this may permit maximum growth and reproductive rates. According to Hamilton (1971, 1973, 1975), their black colouration enables desert beetles to maximize absorption of solar radiation during the early morning and late afternoon, thereby extending the term for surface activity and foraging. This hypothesis is supported by observations on the times of surface activity of sympatric black and white beetles in the Namib desert; and also by the fact that the black *Onymacris plana* (Fig. 32a) exhibits elevated elytron transmittance to short wavelength infrared radiation, which predominates at times when the angle of the sun is low (Henwood 1975). Visible and ultraviolet radiation, which approach their maximum at midday, in contrast, are absorbed by the elytra. The high temperature, in excess of 60 °C reached by the elytron, is claimed to provide additional thermal benefits by reversing the thermal gradient, thereby increasing convective and radiant heat loss (Hadley 1979a).

The maxithermy hypothesis has been criticized by Heinrich (1977) who questions why an animal's biochemical machinery should not have evolved to perform equally efficiently at lower temperatures. At the same time, it has been argued that the function of black colouration in desert animals is aposematic except when it occurs in substrate races that inhabit lava fields and regions where the soil is black. In such places, it may be cryptic (Sect. 7.4.1). In the case of Tenebrionidae, Scarabaeidae and other black beetles, black colouration is, additionally, an evolutionary legacy (Cloudsley-Thompson 1979a). Although this conclusion would appear to be generally applicable, Seely (1985) claims that the tenebrionids of the Namib, unlike those elsewhere which possess defensive secretions (Tschinkel 1975), are both conspicuous and non-toxic. They are consumed in large numbers by arachnids, reptiles, birds and predatory mammals. Seely suggests that, in this instance, the selective value of predation has not been as marked as the selective value of the desert environment, and that the conspicuousness of the fauna is a result of an evolutionary legacy reinforced by selection for physiological adaptations. These include thermal properties (Hamilton 1973): in the Namib desert therefore, the concept of maxithermy could perhaps have an application which it apparently lacks elsewhere (Cloudsley-Thompson 1979a).

Furthermore, Heinrich's (1977) argument fails to explain why black Namib desert beetles, whose food is available at all times of the day and night, are not nocturnal and have not evolved the necessary biochemical adaptations to be active at cool night temperatures. Likewise, Seely et al. (1988b) suggested that the high body temperatures of these beetles reflect both the ease with which small day-active invertebrates withstand temperatures higher than the ambient, and the difficulty of attaining lower temperatures without utilizing water for evaporative cooling. But, again, this does not explain why such beetles are not nocturnal to begin with. More promising is Seely's (1985) earlier hypothesis that high body temperatures have been selected in Namib tenebrionids because, by being day-active, many nocturnal vertebrate predators, which cannot withstand such high temperatures, are avoided. Furthermore, as their metabolic rates increase, the beetles become faster and more adept at escaping their enemies.

Although day-active desert reptiles likewise tend to maintain a high eccritic Tb, this does not necessarily support the maxithermy hypothesis. However, the proposition has not often been tested. Pietruszka (1987), nevertheless, made simultaneous measurements of sand surface temperatures and the occupation of microsites by the Namib dune lizard, *Angolosaurus skoogi*, and compared them with expectations from the maxithermy hypothesis of thermally directed patterns of habitat use. Little correspondence was found between the two, suggesting that *A. skoogi* does not follow a maxithermal strategy of thermoregulation. Rather, it appears to use the sand dune habitat mainly for reasons other than thermal considerations, such as foraging, social interactions and the surveillance of predators. The presence of an immediate thermal refuge in the form of cool substrate sand is probably the principal underlying factor.

4.3.3 Cold Tolerance

In many deserts cold is a limiting factor, and adjustments such as freezing resistance and supercooling, typically associated with high latitude species, have been demonstrated in a number of desert organisms (Hadley 1979b). The responses of desert arthropods and reptiles to cold have been studied far less than their reactions to heat. Nevertheless, cold can also be a limiting factor both at night and during the winter, as we have already seen in the case of the carabid beetle *Thermophilum sexmaculatum* (Sect. 3.4.1). More spectacular adjustments to cold, such as resistance to freezing and supercooling, typical of high latitude and high altitude species, have also been demonstrated in a number of desert forms. Cloudsley-Thompson and Crawford (1970b) obtained an average supercooling point of -3.1 °C from large *Scolopendra polymorpha* (Fig. 14a) collected in summer, while Crawford et al. (1975) found that the same species showed good recovery after supercooling (Sect. 6.4.1). Lowe et al. (1967) found that *Drosophila nigrospiracula* is capable of supercooling to between -5 and -6 °C, a feature which permits these insects to remain active on the surface and breed throughout the winter months in the Sonoran desert; while supercooling points of -11.8 °C have been recorded for the tenebrionid beetle *Eleodes* sp. and -5.8 °C for the scorpion *Vejovis* sp. (Cloudsley-Thompson and Crawford 1970a). Subsequent investigations have provided supercooling values ranging from -3.8 °C in adult desert locusts (*Schistocerca gregaria*) to -10.4 °C in *Trachyderma hispida* (Cloudsley-Thompson 1973).

Lee and Baust (1985) concluded that low temperature acclimation in the North American desert spider *Agelenopsis aperta* (Agelenidae) does not result in depression of the supercooling point, although this may vary seasonally in some spiders, possibly in response to changing photoperiod. Selection of protected overwintering habitats obviate the need for seasonal mechanisms enhancing cold tolerance. The combination of a normal, non-coldhardened supercooling point of -8 to -10 °C and movement into a moderate thermally buffered hibernaculum is likely to provide sufficient protection for this species.

Crawford and Riddle (1975) investigated the overwintering physiology of the semi-montane scorpion *Diplocentrus spitzeri* from New Mexico, and noted increased levels of cryoprotectants such as glycerol and sorbitol in the haemolymph. A substantial seasonal depression of the supercooling point of the scorpion *Paruroctonus aquilonalis*, which enables it to withstand freezing, is apparently associated with the cessation of freezing at the onset of winter (Riddle and Pugach 1976).

Supercooling points for 45 reptilian species, including many desert dwellers, were recorded by Lowe et al. (1971). Mean supercooling limits for lizards and snakes ranged between -3.89 and -7.44 °C. Ability to supercool at this level would provide protection against freezing likely to be experienced by lizards such as *Sceloporus jarrovi* that overwinter in hollow rock crevices.

For some arthropod and reptilian species, the ability to supercool may provide protection against freezing but, for others, the capacity to supercool is greater than the necessity to survive climatic extremes in a particular desert region, and may therefore be a taxonomic rather than an adaptive feature (Cloudsley-Thompson 1973).

4.4 Thermal Control

4.4.1 Metabolic Effects

The periodicity and extent of hyperthermy in arthropods and reptiles and dependent upon the animal's size, metabolism and state of hydration. This adaption is of particular value in birds and mammals whose surface to volume ratio is low and heat storage potential high. A few large insects are known to regulate Tb endothermally, but this is only possible because of the enormous rate of contraction and relaxation of the thoracic muscles during flight (May 1985). Such insects include locusts, moths, bumblebees, honeybees and dragonflies. Concurrently, heat loss from the thorax is frequently restricted by the presence of insulating scales, hairs or subcuticular air sacs which, with the assistance of circulatory mechanisms, make a high Tb at low air temperatures possible.

Pre-flight warm-up by sun basking and shivering of the flight muscles frequently takes place on cold mornings. In contrast, in the heat of the desert, problems of overheating and desiccation are magnified during flight, particularly in large insects such as locusts. Obligate endothermy, resulting from flight metabolism, may raise thoracic temperature more than 30 °C above the ambient air temperature in insects weighing more than about 0.5 g (Heinrich 1975). Evaporation, which may remove 8-9% of the total heat produced in the thorax, will be discussed in Section 5.2.3.1. In small insects, heat loss by evaporation is usually small and unregulated.

These effects are enhanced by the concentration and numbers of social insects (Sect. 3.5) but, only in mound-building termites, can they be regarded as being adaptations to desert conditions. In the case of reptiles, however, which are mostly much larger than arthropods, metabolic heat production and evaporative heat loss are sometimes important.

In their natural environments, animals are exposed to a number of energy-exchange processes. In addition to gain and loss of energy by conduction, convection and radiation, reptiles produce metabolic heat which may be conserved by shunting blood from superficial to deep tissues and by decreasing the heart rate (see Cloudsley-Thompson 1971; Bartholomew 1982; Huey 1982;

Tracey 1982; etc.). A fine example is afforded by Heath's (1965, 1966) classic study of the American horned lizard *Phrynosoma coronatum* (Fig. 39). This species, like *Amphibolurus ornatus* (Sect. 4.3.1), shows a complex pattern of interrelationships in the thermal regulation of physiological and behavioural responses to temperature (Fig. 40). Emergence from the sand during hours of normal activity depends upon a high head temperature and is independent of the temperature of the body, but the differential diminishes at elevated Tb. Moreover, head temperature in this species is regulated by opening lateral shunt vessels from the cephalic sinuses to the external jugular veins, which bypasses a countercurrent heat exchange between the internal jugular veins and the internal carotid arteries (Fig. 41).

Comparable observations have since been made by other investigators. For instance, Riedesel et al. (1971) found that, whereas the heart rate of the box terrapin *Terrapene ornata* is correlated with Tb rather than with the temperature of the head, thermoregulatory salivation (Sect. 5.2.3.2) may depend upon both. Blood from the body, however, may influence brain temperature before coming into equilibrium with the temperature at the surface of the head. Campbell (1969) showed that the temperature of maximum auditory sensitivity in six species of lizard varied according to the preferred temperature of the species.

Fraser (1985) found that heating rates in *Egernia cunninghami* are significantly greater than cooling rates. In addition, faster heating and cooling rates were recorded during the summer than in winter. The ability to regulate the flow of heat within the body enables reptiles to enjoy a certain amount of flexibility in the exchange of thermal energy with the environment, and even the smallest desert lizards have the capacity to control their rates of heat exchange with the

Fig. 39. Horned lizard, *Phrynosoma coronatum*, New Mexico

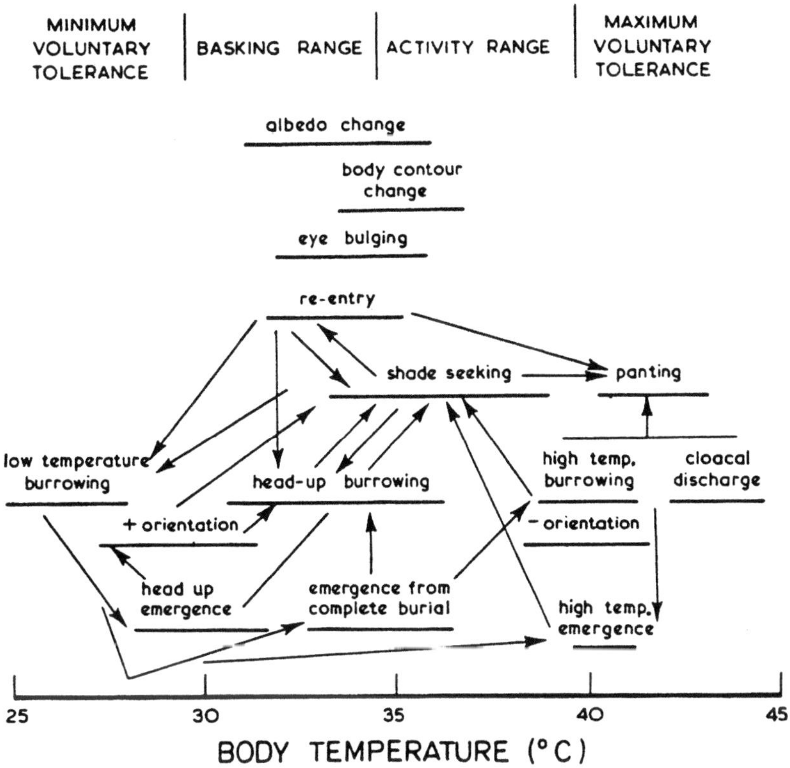

Fig. 40. The interrelations of temperature-regulating behaviour in *Phrynosoma coronatum*. The range of body temperature for each pattern is given (Cloudsley-Thompson 1971 after Heath 1956)

environment (McKenna and Packard 1975). Templeton (1960, 1970) showed that, as the environmental temperature increased, the oxygen consumption and, thereby, the metabolic heat production of *Dipsosaurus dorsalis* also increased. He suggested that respiratory cooling does not serve to lower a lizard's body temperature but, rather, to dissipate the increased metabolic heat engendered. In some tests he wrapped lizards in insulating material, and found that the production of metabolic heat could then be lethal.

According to Whitfield and Livezey (1973), lizards fall into three groups with respect to their tolerance of high temperatures. The most highly thermophilic desert species, with a CTM between 47.1 and 48.7 °C, show thermoregulatory patterns which reduce their rate of heating or even lower the body temperature from 0.1-5 °C. These temperature control mechanisms seem to be physiological and include panting and cloacal discharge (Sect. 5.2.3.2).

Bennett and Dawson (1972) determined the oxygen consumption and lactate content of *Dipsosaurus dorsalis* under standard conditions and after a bout of maximum activity. They remarked on the degree of energy mobilization during,

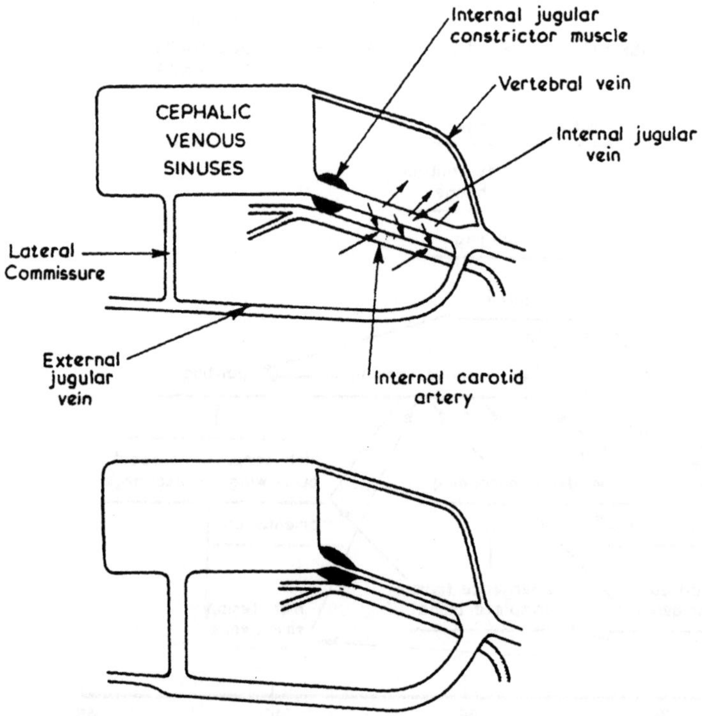

Fig. 41. Diagram showing the relations of the major vessels in the head of the horned lizard. The venous spaces of the head are abstracted to a single space. *Above* Internal jugular vein is open. *Arrows* indicate heat exchange between internal jugular veins, internal carotid artery and tissues of the neck. *Below* Closure of the internal jugular constrictor muscle causes collapse of the internal jugular vein. Venous blood returns through a shunt to the external jugular vein. Cool blood, warmed only by the neck tissue, enters the head, while warm blood flows through the external jugular vein to the body. (Cloudsley-Thompson 1971 after Heath 1966)

and rapid recovery after, activity, and postulated that these may represent the principal physiological adaptations of these reptiles to their preferred thermal levels. Bennett and Gleeson (1976) measured oxygen consumption and lactate production during and after bursts of activity in *Sceloporus occidentalis*. Maximum levels of oxygen consumption were achieved at 30-40 °C. The maximum aerobic increase above resting levels occurred at 35 °C, the preferred body temperature. Combination of energy utilization, in both aerobic and anaerobic modes, gives a highest activity capacity at 30-35 °C, the range of body temperature normally experienced during daytime throughout the year.

Thermoregulatory behaviour has been studied more often in desert lizards (Dawson 1967; Cloudsley-Thompson 1971; Bartholomew 1982) than in desert snakes. Analysis of heating and cooling rates, metabolism and the ratio of heart rate to body temperature in the crepuscular thigmothermic colubrid *Spalerosophis cliffordi* of the Negev desert indicates some capacity for physiological control: live snakes heated more rapidly than they cooled and the mean rate of cooling was 82% that of heating. Heart beats were at their maximum rate, and metabolism near the minimal level during heating. While cooling, metabolism was maximal and heart rate was minimal (Dmi'el and Borut 1972). Thermogenesis by shivering has long been known to occur in pythons while brooding their eggs (Hutchison et al. 1966). It has recently been analyzed in the Australian *Morelia (=Python) spilota*, in which the energetic costs are so high that, in cool temperate regions, breeding may occur only every second year (Harlow and Grigg 1984). Indeed, it may well occur in other large Boidae although it is probably uneconomical in smaller species such as the Australian *Liasis childreni* (Fig. 42). These are believed to shiver while brooding their eggs, but have not been shown conclusively to increase Tb by this mechanism.

By means of miniature temperature-sensitive radiometers, surgically inplanted into free-ranging adult diamond pythons, *Morelia spilota*, Slip and Shine (1988) showed that brooding females maintained a body temperature differential above ambient of about 9 °C and occasionally up to 13 °C. Their temperatures were significantly higher and less variable than those of non-brooding females or of males. The energetic costs of brooding are indeed high, but doubtless outweighed by the benefits of rapid embryonic development and high embryonic survival rate. Various hormones have been suggested to play a role in reptilian thermoregulation but there is little direct evidence in favour of this. The subject

Fig. 42. *Liasis childreni* (Boidae), Central Australia

has been reviewed by Bradshaw (1986) and will not be discussed in this book. The primary physiological processes involved in thermoregulation of desert reptiles are: (1) increased heat production through metabolism; (2) variations in the rate of heat exchange with the environment through circulating adjustments and changes in dermal vascularity; and (3) systematic changes in colour and reflectivity of the skin (Sect. 4.4.2). The fourth physiological thermoregulatory process is evaporative cooling, which will be discussed later (Sect. 5.2.3).

4.4.2 Colour and Reflectivity

Dark-coloured bodies heat at a faster rate than light-coloured objects because reflectance is lower and the rate of absorption of radiant energy faster. Nevertheless, in the case of arthropods, the thermal effects of colour are probably comparatively slight (Edney 1971a; Cloudsley-Thompson 1979a), despite the claims of Hamilton (1973, 1975). Although Bolwig (1957) found somewhat higher temperatures in the subelytron space of the black tenebrionid *Onymacris multistriata* than in the white *O. bicolor*, the black species was cooler at Ta above 38 °C, perhaps because of ventilation. On the other hand, Hadley (1970a) observed a fairly constant depression of about 4 °C when he coloured the black elytra of *Eleodes armata* with white paint and exposed both experimental and control animals to the desert sunshine. Edney (1971a) obtained a similar differential in sunlight between the temperatures in the subelytron cavities of black *Onymacris rugatipennis* and white *O. brincki* in the Namib desert: he concluded that the selective advantage of white elytra probably has little to do with heat balance.

Interference (metallic) colours, likewise, are not effective in reducing heat gain from solar radiation, although beetles with white elytra probably experience significantly lower body temperatures than do either black or iridescent beetles under comparable environmental conditions, according to Schultz and Hadley (1987). This conclusion was based on a comparison of reflectivities and heat gain of cuticles from brightly metallic and achromatic forms of two polymorphic tiger beetle species, *Cicindela horni* and *C. formosa* in the southwestern United States. The authors, nevertheless, conceded that there was strong evidence that the expanded maculations of some Cicidelidae are important in matching white backgrounds and thus avoiding predators.

In a recent experimental study, Turner and Lombard (1990) concluded that body colour affects only one avenue of heat exchange — the absorption of direct shortwave and visible radiation. It is irrelevant to all other avenues of heat exchanges, namely convection heat loss, reflected shortwave and visible radiation, and heat from the ground. Of all the factors that affect beetle temperature, convection heat loss is the most important. Of lesser importance, but still significant, are ground temperature, ground colour and incident visible radiation. Longwave radiation from the ground is a negligible component of a beetle's exchanges of heat. Under semi-nocturnal conditions, these factors interact to render colour a negligible influence on the body temperatures of *O. bicolor* and *O. unguicularis* despite their stark differences in body hues.

The experimental evidence in support of a thermal significance in the black colouration of many desert insects is therefore contradictory and, at best, inconclusive (Sect. 4.3.2). Colour change, in response to high Ta, does not appear to take place, even among the few terrestrial arthropods that are endowed with the ability to change their hues and maintain crypsis against differently coloured backgrounds.

On the other hand, although colour change, which is not uncommon in reptiles, may also be concerned mainly with concealment (Cloudsley-Thompson 1971), it also plays some part in thermoregulation. Atsatt (1939) observed different lizards over a range of Ta and noted that their colours were lighter at higher Tb. He concluded that these colour changes facilitated temperature regulation. Similar assumptions have been proposed and elaborated by numerous subsequent authors. Cowles (1958), however, was the first to suggest that changes in the flow of the subcutaneous blood might also be involved. Studies on heating and cooling rates in lizards, in which marked hysteresis has been found, point to the existence of temperature-dependent changes in peripheral blood flow (Bartholomew and Tucker 1963, 1964; Weathers 1970). Weathers (1971) found that local changes in subcutaneous blood flow in *Dipsosaurus dorsalis* are a direct result of heat on vascular smooth muscle and Baker et al. (1972) suggested that the peripheral response to local heating is dependent on the thermal load, and that changes in peripheral patterns are independent of heart rate and deep body temperature in *Iguana iguana* and *Tupinembis nigropuretatus*. A linear increase in reflectivity with Tb was demonstrated experimentally in *Liolaemus multiformis* by Pearson (1977), while Rice and Bradshaw (1980) found the same in *Amphibolurus nuchalis*. Subcutaneous blood flow and skin reflectivity varied directly with Tb: peripheral blood vessels dilated during basking as a result of local heating, facilitating an increase in dermal blood flow. In this way the effects of behavioural thermoregulation are reinforced by physiological control mechanisms. Regal (1975) argued that the enlarged scales of desert lizards may act as shields to radiation and provide protection from overheating.

The lizard *Liolaemus multiformis*, which lives at high altitudes in the arid Peruvian Andes, basks on insulating surfaces instead of on rocks, and thereby warms at 0.6 °C min^{-1} faster. The skin blanches as Tb increases, and the reflectance of red light also increases by about 0.8% per °C (Pearson 1977). Surprisingly, Fraser (1985) found that when the Australian scincid lizard *Egernaria cunninghami* was painted black, its rate of radiant heating was retarded rather than enhanced.

Colour changes in reptiles are correlated both with concealment (Sect. 7.4.1) and with thermoregulation. The ability to reflect light is more marked in desert lizards than in species from other biomes, suggesting that colouration does have a thermal influence, but the operation of concealing and thermoregulatory colouration are largely synergistic. When the two tend to produce opposing effects, the one possessing the greater survival value is selected (Cloudsley-Thompson 1979a).

4.5 Responses to Ultraviolet Radiation

The damaging effects of ultraviolet light on living organisms have long been known, and day-active desert dwellers are exposed to large doses of solar radiation. Hadley (1985) states that surface waxes increase the reflection of potentially harmful ultraviolet radiation from the cuticles of desert insects, and also delay the rise in their body temperatures whilst foraging.

Although infrared radiation penetrates to some extent both the cuticle of arthropods and the skin of reptiles, the transmission of ultraviolet light is nil. Therefore, this cannot provide an explanation to the black peritonum and pleura of desert reptiles, since protection of the tissues of the reptile body by the corneous part of the integument is extremely efficient against all kinds of solar radiation (Bodenheimer 1953). Nevertheless, it has been demonstrated that, without its black peritoneum, *Dipsosaurus dorsalis* would not be able to change colour, and thereby adapt to the rigours of the hot desert environment, without exposing internal biochemical processes to sufficient visible and near ultraviolet light to produce mutations which might easily affect the maturation of its gonads (Porter and Norris 1969). Ultraviolet light has been shown to influence the behaviour of *Dipsosaurus dorsalis, Crotaphytus collaris* and *Agama agama*, inducing threat displays to conspecifics in *D. dorsalis* and *C. collaris* and an increased frequency of head nods in *A. agama*. Day-active desert lizards are exposed to large amounts of ultraviolet light in nature and have evidently become adapted to it (Moehn 1974).

4.6 The Mechanism of Heat Death

Although a considerable amount of work has been carried out on the lethal temperatures of animals, the physiological mechanisms responsible for the onset of death remain obscure. Coagulation of protein was once a common explanation but, for many animals, lethal temperatures are much lower than the temperature at which proteins begin to coagulate. Lack of oxygen at high temperatures has also been thought to be a primary cause of heat death.

In the case of desert arthropods there is evidence of the accumulation of acidic metabolites in the haemolymph. In the case of the African scorpion *Leiurus quinquestriatus* and the tenebrionid beetle *Trachyderma* (= *Ocnera*) *hispida*, exposure to near-lethal temperatures for 24 h results in marked decreases in the pH of the blood (Cloudsley-Thompson 1962c). Similar results have been obtained by Ahearn (1970a) in the case of the North American desert tenebrionid *Centrioptera muricata*. Both uric acid and lactic acid rose in the haemolymph of animals exposed to lethal temperatures, while the osmotic pressure also increased. In addition, there was a decrease in sodium and an increase in potassium. The similarity between the ionic ratio curve and the LD_{100} curve suggests a correlation between the two factors.

As long ago as 1935, Edwards and Dill suggested that decreased affinity of haemoglobulin for oxygen above 45 °C might be an important factor limiting the survival of reptiles at high temperatures. Other workers have produced supporting evidence (reviewed by Cloudsley-Thompson 1971) but no difference in survival at high temperatures was observed in *Mabuya quinquetaeniata* or *Tarentola annularis* regardless of the oxygen content of the air during exposure (Cloudsley-Thompson 1965a). Furthermore, irreversible damage may occur when normal oxygen consumption can still take place. Again, the degree of unsaturation of phosphatides may sometimes, but not always, be correlated with lethal temperatures. In desert reptiles, whose lethal temperatures lie between 45 and 50 °C, the cause of heat death may well be connected with direct effects on tissue proteins and enzymes (Licht 1968).

4.7 Conclusions

Of the five processes known to influence the body temperatures of animals: radiation, convection, conduction, evaporation and metabolism, the first three are mainly employed in the thermoregulation of both arthropods and reptiles. Metabolic heat production is relatively low, except in the case of large insects in flight, and pythons brooding their eggs, but differential blood shunting allows desert lizards to gain heat more rapidly than they lose it, and thus remain at their eccritic temperature for longer periods than would otherwise be the case. Evaporative cooling is limited. Thermoregulation is primarily behavioural in desert arthropods and reptiles, while adaptations to water shortage tend to be mainly physiological, as will be seen in Chapter 5. There is naturally, however, considerable interaction between physiology and behaviour, thermal regulation and water conservation.

5 Water Balance and Nitrogenous Excretion

5.1 Introduction

All aspects of ecophysiology are interrelated but, in hot deserts, none more closely than thermal and water relations. Because the drying power of the air is so greatly increased at high temperatures, heat and drought are the two most important elements of the climate of arid environments. The problem of dehydration in desert animals reflects both the high saturation deficiency of the atmosphere, which is usually more difficult to avoid than high temperature, and the use of water for evaporative cooling. Water is lost by evaporation through the integument — the cuticle of arthropods and the skin of reptiles — in respiration and thermoregulation, with the faeces, in nitrogenous excretion, and in the secretion of toxic repellents. It is obtained through drinking, by ingestion of moist food or soil, absorption from the air and as a product of metabolism (Wharton 1987). The water relations of desert arthropods have been reviewed among others by Cloudsley-Thompson (1964a, 1975), Crawford (1981), Edney (1974, 1977) and of reptiles by Cloudsley-Thompson (1971, 1988b), Mayhew (1968), Schmidt-Nielsen (1964), Schmidt-Nielsen and Dawson (1964).

5.2 Evaporative Water Loss

The major source of water loss from desert arthropods is probably through cuticular transpiration, less in respiration except in the case of large insects when in flight (Sect. 5.2.3.1), and very little in the faeces or through nitrogenous excretion. Because they are larger than arthropods, most desert reptiles can afford higher rates of integumental transpiration and respiratory and excretory water loss.

5.2.1 Integumental Transpiration

Integumental transpiration usually predominates at lower ambient temperatures, respiratory water loss at higher temperatures when evaporative water loss is invoked for thermoregulatory purposes. In the case of North American tenebrionid beetles, such as *Eleodes armata* and *Cryptoglossa verrucosa*, a greater percentage of water is also lost via quinone secretions (Sect. 7.5.2) at higher temperatures and humidities (Ahearn and Hadley 1969). Many arthropods from temperate regions respond promptly to dehydration by moving into a damper atmosphere, but desert species are far less sensitive to water loss, and only seek a moist environment when severely desiccated (El Rayah 1970b, Andersen et al. 1986).

5.2.1.1 Arthropods

In most desert-adapted arthropods and reptiles, the rates of water loss through transpiration are low compared with rates from non-desert species. In the Arachnida and Insecta, an epicuticular layer of wax, as well as the sclerotized chitin-protein complex of the exocuticule and extensions of lipid material into the underlying cuticular layers, are responsible (Edney 1977; Hadley 1979a,c). Even desert woodlice and millipedes, however, which lack an epicuticular wax layer, exhibit extremely low rates of water loss through transpiration (Crawford 1972; Cloudsley-Thompson 1977c) and this is undoubtedly adaptive. For instance, compared with the mesic *Oniscus asellus*, which transpires at a rate of about 130 $\mu g\ cm^{-2}\ h^{-1}\ mmHg^{-1}$ at 30 °, and a rate of 110 mg from *Porcellio scaber*, which also inhabits mesic environments, the North American desert woodlouse *Venezillo arizonicus* (Armadillidae) loses only about 32 mg $h^{-1}\ mmHg^{-1}$, the Great Palaearctic desert species *Hemilepistus reaumuri* (Porcellionidae) 23, and the Arabian *Porcellio evansi* as little as 20 (Constantinou and Cloudsley-Thompson 1987). Figures from desert insects under comparable conditions include *Arenivaga investigata* (Dictyoptera) 12, *Cryptoglossa verrucosa* 8, and *Centrioptera muricata* 6. The last two are Tenebrionidae. From desert arachnids we have rates of 6.6 mg $cm^{-2}\ h^{-1}\ mmHg^{-1}$ from *Galeodes granti* (Solifugae: Galeodidae), 3.6 from the theraphosid spider *Eurypelma* sp., while figures from scorpions include *Hadrurus arizonensis* (Vejovidae) 1.22, *Buthotus minax* (Buthidae) 0.98 and *Androctonus australis* (Buthidae) 0.8 (most of these figures are cited by Edney 1977). Cuticular and respiratory transpiration constitute the two main avenues of water loss from *H. arizonensis*, cuticular water loss predominating at temperatures up to about 38 °C and respiratory transpiration above about 40 °C (Hadley 1970b, 1990).

Despite the number and diversity of arthropodan families, the integumental plan is basically similar in all. As with the cuticle of plants, the arthropod cuticle is waterproofed by impervious layers of wax (see Hadley 1981, 1984, 1989; Cloudsley-Thompson 1988a). Surface wax deposits also occur in many desert insects and arachnids, while elaborate multicoloured wax blooms are often present in desert insects. Doubtlessly they complement the role played by epicuticular lipids in restricting water loss. For instance, the banded grasshopper *Trimerotropis pallidipennis*, which experiences air temperatures in excess of 50 °C, is uniformly covered by wax encrustations, crystals and filaments. The most intricate wax blooms, however, are found in certain desert Tenebrionidae. *Cryptoglossa verrucosa*, a common beetle of the Sonoran desert, exhibits distinct colour phases that range from light blue to jet black when subjected to low and high humidity extremes respectively (Hadley 1985). The same effect has been described in other desert species.

The morphological basis for these colour phases are wax filaments that radiate from the tips of miniature tubercles to cover the integument at low humidity (Hadley 1979c). McClain et al. (1984) monitored the dynamics of the extracuticular wax bloom of *Cauricara phalangium* in the Namib desert. They found that rates of water loss were significantly lower from individuals collected in May, when a full bloom was present, than from those taken in August, when little or no wax bloom was present. They concluded that the wax bloom contributes to the protection of these day-active beetles against high temperatures and radiant heat loads. In the Namib, wax blooms occur in more than 50% of beetle species of the tribes Adesmiinae and Zophosini. The percentage with a wax bloom increases from the coastal fog desert to the inland hot dry desert (McClain et al. 1985). The dermal glands concerned with its production have been described by Hanrahan et al. (1984). Several functions have been demonstrated for wax blooms. In addition to reducing water loss, they include protection from predators, microorganisms, ultraviolet light and mechanical abrasion by the substrate. They also increase reflectance of solar radiation and serve in chemical communication (Hadley 1984).

Although the epicuticular wax layer is protected from abrasion in most desert arthropods, this is not invariably the case. Adult vinegaroons or whip scorpions, *Mastigoproctus giganteus* (Uropygi; Fig. 43), readily lose water to the air at 26 °C at relative humidities up to 95%. As temperatures increase, they lose water in proportion to the saturation deficit of the atmosphere (Fig. 44), suggesting the absence of an epicuticular wax layer (Crawford and Cloudsley-Thompson 1971). This, however, is not the case: Ahearn (1970b) found that young specimens show a typical 'critical' temperature (37.5 °C) above which water is lost rapidly through transpiration (Fig. 45). The answer can only be that continuous contact with the sand by older specimens must result in abrasion of the waterproofing epicuticular layer of wax (Crawford and Cloudsley-Thompson 1971; Cloudsley-Thompson 1988a). Why *M. giganteus* should be so vulnerable to abrasion, when scorpions and other arachnids are not, is difficult to understand.

Several layers of the arthropod cuticle participate actively in the waterproofing process. The principal barrier to water loss is a layer or layers of lipids associated with the epicuticle. In addition, the chitin-protein complex of the endocuticle and even the epidermal cells that secrete and maintain the cuticle may contribute to its impermeability. The chemical composition of the epicuticular lipids has been correlated with cuticular permeability in a number of desert insects and arachnids (reviewed in Hadley 1982, 1985). For example, the lower cuticular permeability of the Sonoran desert scorpion *Hadrurus arizonensis*, when compared with the mountain species *Uroctonus apacheanus*, can be attributed in part to the higher surface densities of lipids and hydrocarbons in the former (Toolson and Hadley 1977).

Seasonal differences in permeability are also linked to the composition of cuticular lipids. Rates of water loss from the scorpion *Centruroides sculpturatus*

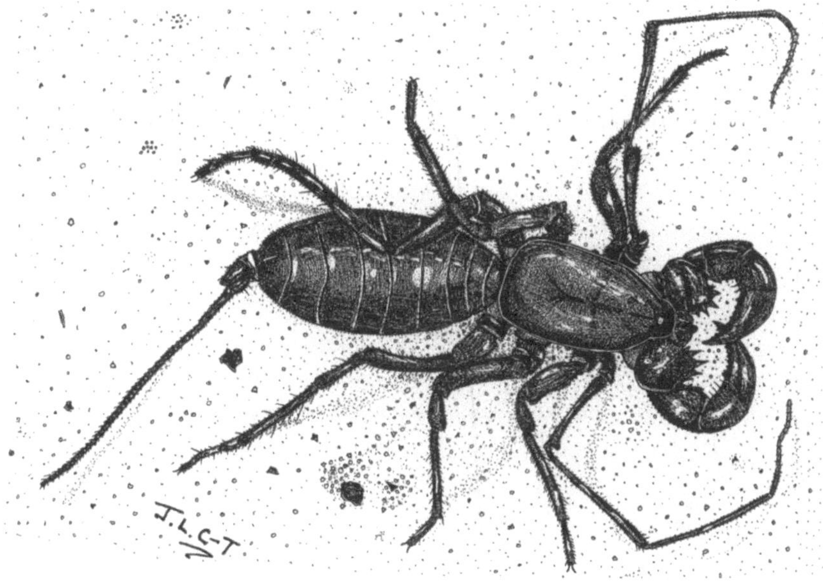

Fig. 43. *Mastigoproctus giganteus* (Uropygi), North American desert

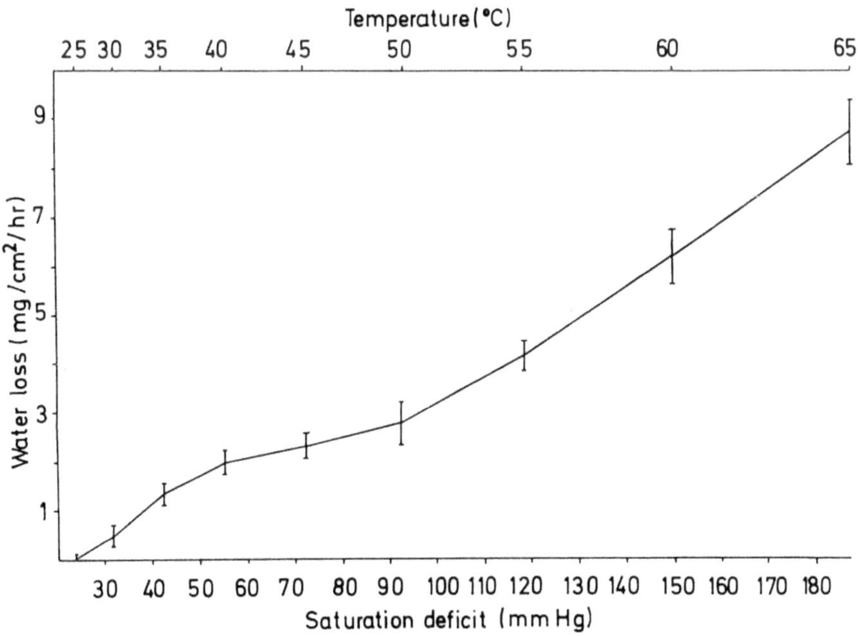

Fig. 44. Transpiration of *M. giganteus* in relation to the temperature of the ambient air. Each point represents the mean hourly water loss (1 standard error) of ten animals (Crawford and Cloudsley-Thompson 1971)

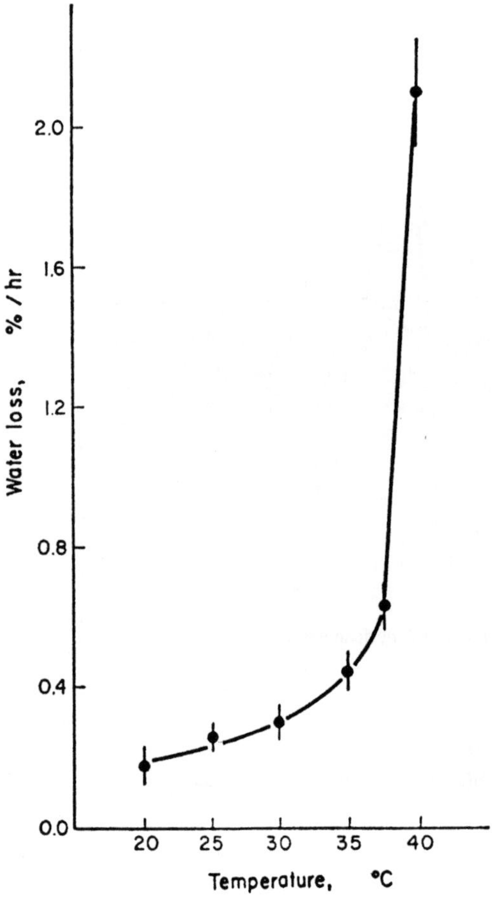

Fig. 45. Water loss by *M. giganteus* as a function of temperature. Each *point* represents the mean water loss (expressed as percentage initial body wt/h) of at least nine individuals exposed for 2 h to selected temperatures at approximately 0% RH in slowly moving air. The *vertical lines* are one standard deviation of each mean (Ahearn 1970b)

are significantly lower in summer than in winter. This can probably be accounted for by an increased percentage of long-chain epicuticular hydrocarbons in summer (Toolson and Hadley 1979; Sect. 6.4.1). Again, *Eleodes armata* (Tenebrionidae) in summer, or in winter after acclimation to 35 °C for 5-10 weeks, exhibited a higher percentage of long-chain hydrocarbons than did normal beetles in winter or appropriate controls (Hadley 1977).

Regional differences in cuticular permeability have been found to occur in the North American desert cicada *Diceroprocta apache*. Hadley et al. (1989) measured transpiration in vivo from three regions of the dorsal surface. Transcuticular water flux rates were similar for all three at 27 °C but, at 41.5 °C, rates increased sharply from the dorsal thorax and abdomen — regions that contain large pores (7-8 mm in diameter) — but not from the lateral thorax which does not contain such pores. Active extrusion of water through the pores began at 39.2-39.3 °C, the temperature at which cicadas normally seek cooler microclimates.

The function of the subelytron cavity in tenebrionid beetles has been the subject of some speculation (Fiori 1977). As a result of a preliminary experimental investigation, the suggestion was made that it serves to reduce transpiratory water loss (Cloudsley-Thompson 1964b). This conclusion has been substantiated by Ahearn and Hadley (1969) and Ahearn (1970c). When the elytra are removed, exposing the subelytron cavity and the dorsum of the abdomen, the rate of transpiration increases. Edney (1971a) and Hadley (1972) also suggested that, since the subelytron cavity is filled with air, it may act as a buffer zone, retarding the flow of air from the elytra to the abdomen. In the case of the North American *Eleodes obscura*, Slobodchikoff and Wismann (1981) established a significant correlation between water loss and the depth of the subelytron chamber. This implies that the chamber may be a structure that permits the rapid expansion of the abdomen, providing the beetles with a mechanism by which they can rapidly drink large quantities of water from ephemeral sources such as rain puddles. As the beetles drink, the abdomen expands rapidly and the chamber shrinks. Storage of water in this way is probably of particular importance to species that spray or exude defensive secretions when attacked by predators.

Although all millipedes (Diplopoda) lack a discrete waterproofing epicuticular wax layer, as is found in insects and arachnids, and therefore lose water more quickly by transpiration, some species transpire sufficiently slowly to be able to exist in desert regions, as mentioned above. An example is afforded by *Orthoporus ornatus* (Spirostreptidae) in the North American deserts (Crawford 1972), another by *Harpagophora nigra* of the Namib desert. In the latter, water balance during winter seems to depend partly on the ingestion of detritus and moist soil (Sect. 5.6.2). Although resistance to desiccation is not exceptional for a desert millipede, individuals losing up to half their body weight can quickly regain water by oral uptake from damp surfaces (Crawford and McClain 1983).

Centipedes show even less physiological control of cuticular transpiration than do millipedes, but some Scolopendromorpha, such as *Scolopendra polymorpha* (Fig. 14a) of the Chihuahuan desert, are able to survive through their nocturnal habits (Cloudsley-Thompson and Crawford 1970b), while larger species, such as *S. heros*, may even be seen moving freely in the open during daylight hours (Crawford 1981).

5.2.1.2 Reptiles

Despite the fact that reptiles lack sweat glands, their rates of cutaneous water loss are somewhat higher than previously thought to be the case, albeit lower in desert than in mesic species. In most cases, little is known of the morphological and physiological factors responsible, but the presence of scales is not significant. Rates of water loss from a scaleless gopher snake, *Pituophis melanoleucus catanifer*, were found to be comparable with those from a normal individual (Licht and Bennett 1972), despite the fact that the scaleless snake even lacked the outer superficial dermal layer, and possessed a keratin layer much thinner than normal. Lipids, similar to those in the insect epicuticle, are probably responsible for reduced cutaneous transpiration in reptiles (Hadley 1982).

Many workers have measured evaporative water loss from desert reptiles (e.g. Chew 1961; Cloudsley-Thompson 1965a, 1970a, 1971, 1979b; Dawson et al. 1966; Schmidt-Nielsen and Bentley 1966; Roberts 1968; Minnich 1970a,b; Nagy 1972; Templeton 1972; Shoemaker 1975), but such work does little to indicate the extent of adaptation to desert conditions unless comparisons are made with related species that do not inhabit arid environments. This has been suggested by Gans et al. (1968), Krakauer et al. (1968) and, indirectly, by Mautz (1982), as well as by some of the authors of the publications reviewed by Templeton (1970), Lillywhite and Maderson (1982), and Nagy (1982). This work renders superfluous, further extensive treatment of the subject, apart from the brief mention of a few of the more recent publications that introduce new ideas, concepts and approaches.

Not only are rates of water loss from different species of reptiles highly correlated with the aridity of the habitats in which they live (Shoemaker and Nagy 1977; Mautz 1982; Nagy 1982), but rates of evaporative loss are greater in members of populations of the same species (*Anolis roquet* and *A. marmoratus*) that live in more humid than in drier environments. Moreover, rates of water loss are significantly lower in *A. roquet* after 6 weeks' acclimation to more xeric conditions (Hillman et al. 1979). Again, Kobayashi et al. (1983) found that, when maintained for more than 7 days in dry air (<30% RH), *Anolis carolinensis* had rates of total evaporation less than half those from lizards maintained in humid air (>95% RH). This indicates that these reptiles are able to reduce water loss by evaporation under desiccating conditions, as has previously been observed in other species by Cloudsley-Thompson (1965a) and Warburg (1965a). Evaporation through pieces of isolated skin taken from lizards kept in dry air was less than half that through the skins of lizards of the same species maintained in humid air. These results suggest that some of the low rates of transpiration recorded from desert reptiles may represent immediate adaptive responses rather than inherited characteristics (Sect. 8.5.1).

The long-tailed lizard *Tropidurus peruvianus*, one of the most conspicuous epigean animals of the desert of western South America, has been shown to select habitats according to the environmental humidity. This species is present neither on the drier desert plateaux, nor in the more mesic portions of the *lomas* (vegetational belts that appear in response to the condensation of marine fog on the westward side of the coastal range of mountains) (Sect. 5.6.1; Péfaur and Lopez-Tejeda 1983).

Many burrowing lizards have nasal valves which prevent sand grains from being inhaled (Stebbins 1943, 1948). Once a pack of moist sand has formed over the nostrils, it may reduce respiratory water loss by humidifying inhaled air and trapping some of the water vapour from the exhaled air (Pough 1969).

Evaporative water loss from the amphisbaenian *Diplometopon zarudnyi* of Kuwait and Saudi Arabia is surprisingly low, in view of its subterranean mode of life (Gans et al. 1968; Cloudsley-Thompson 1979b). It seems probable that, here again, the major avenue of water loss in this animal is cutaneous rather than respiratory, as it is in the South American *Amphisbaena alba* (Cloudsley-Thompson 1977d). While at rest, the Negev desert snake *Spalerosophis cliffordi* is able to maintain a positive water balance even when no drinking water is available (Sect. 5.6.1). This amount, 2.25 mg g^{-1} day^{-1}, is sufficient to cover the needs of the reptile during activity (Dmi'el and Zilber 1971). The main source of water loss in snakes, as in other reptiles, is through evaporation, and respiratory water loss is often only a small fraction of the total transpiration.

Since body temperature has a large effect on the rate of energy metabolism in ectothermic animals, it is a reasonable supposition that nocturnal lizards should have lower daily energy requirements, and conserve water to a greater extent, than do day-active lizards. Nagy and Degen (1988) tested this hypothesis on the nocturnal gecko *Ptyodactylus hasselquistii* of the Negev. Using doubly labelled water ($HT^{18}O$), they measured both field metabolic and water influx rates. To their surprise, they detected no difference between these and values calculated for diurnal iguanid lizards occupying otherwise comparable ecological niches (Sect. 5.4.2). They suspected that the primary reason why the geckos had such high field metabolic rates might be because they selected rather high body temperatures, especially at the beginning of the day when they were inactive. The hypothesis, therefore, may remain valid for other desert reptiles.

5.2.2 Transpiration from Eggs

Due to their small size, the eggs of all terrestrial arthropods, except those deposited in very humid places must, of necessity, be extremely resistant to water loss. The diversity of terrestrial arthropods is reflected in the diversity of their eggs (Hinton 1981). Embryonic membranes which, in vertebrate amniotic eggs, protect the embryos from adhesion and desiccation, serve the same function in the eggs of arthropods. Their enhanced resistance to osmotic rupture, desiccation and drowning may well explain the low incidence of post-zygotic parental investment among insects, compared with other terrestrial arthropods in which such resistance is less marked. Parental care is common in Arachnida (Sect. 3.5). Only Araneae and Acari habitually lay their eggs in exposed sites and, in both these classes, accessory structures, which afford protection, are usually present. Spider eggs are protected by silken egg sacs, while most of the mites that are also oviparous, protect their eggs in a variety of other ways (Zeh and Zeh 1989).

Millipedes of several different families coat their eggs with pellets of chewed litter, excrement or a mixture of earth and excrement (Cloudsley-Thompson 1958). In some species, however, only a secretion from the maternal reproductive tract covers the eggs. The function of egg pellets is not well understood. Crawford and Matlach (1979) have shown in the case of *Orthoporus ornatus* that the pellets provide a degree of moisture stability for the developing eggs. After hatching, the pre-emergence larvae imbibe water from moist substrates both orally and through the rectum. Since the soil of the pellet is often relatively dry, Crawford and Matlach (1979) concluded that a more consistently beneficial advantage is that of assisting osmoregulation in the larvae. In this respect, pre-emergence larvae of *O. ornatus* resemble older instars of the same species (Riddle et al. 1976). It is quite probable, too, that coating the eggs with pellets may protect them from predation. This trait is by no means confined to desert-inhabiting millipedes but is probably especially valuable in arid environments.

The structure of egg shells and the water relations of reptilian eggs have been reviewed by Packard et al. (1977, 1982). In general, eggs that develop in humid surroundings tend to give rise to larger hatchlings than do eggs which develop under drier conditions. Consequently, there is probably selection pressure on gravid females to locate their nests in sites of optimum water potential. Furthermore, when the thermal capacity of the substrate is great, reptile eggs take up more water under wet conditions and lose less under dry conditions (Ackerman et al. 1985).

Environmental adaptations of reptile eggs have been little studied, but it is clear eggs should not be considered as passive systems responding solely to changes in their physical environment. Thus, Muth (1981) has shown that the eggs of *Dipsosaurus dorsalis* may simultaneously transpire and absorb water, while maintaining a net water potential below the ambient water potential. Temporal changes in the mass of eggs of the zebra-tailed lizard *Callisaurus draconoides* have been reported by Packard et al. (1981), and in those of *Agama persica* by Cloudsley-Thompson (1979b). It would appear that the eggs of desert lizards are generally sensitive to desiccation and, therefore, must be laid at the time of rainfall if they are to survive. The soil environment imposes constraints on the timing of oviposition and the location of suitable sites for egg burrows of *Dipsosaurus dorsalis*. Eggs hatch normally between 28 and 38 °C at environmental water potentials between -50 and -1500 kPa. Egg burrows are therefore located at depths above 22 cm in washes and sparsely vegetated areas away from creosote bushes (Muth 1980).

5.2.3 Thermoregulatory Heat Loss

Once ambient temperature reaches or exceeds body temperature, evaporative cooling is the only means available for dissipating heat. Because of their large surface to volume ratio, small animals such as arthropods and reptiles tend to absorb more heat than do larger ones. Consequently, they cannot afford to use water for thermoregulation except in exceptional circumstances.

5.2.3.1 Flying Insects

The thoracic muscle of insects generate a considerable amount of heat during flight. In small insects the rate of convective heat loss is so rapid that little of this heat is retained in the muscles, and the thoracic temperature probably does not increase more than about 1 °C above the ambient temperature. In larger forms, however, such as locusts, bees and moths, obligatory endothermy may raise thoracic temperature more than 30 °C above the ambient (Sect. 4.4.1). To maintain a high metabolic rate, the flight muscles must be continuously supplied with oxygen which is absorbed through the moist walls of the tracheoles. In consequence, ventilation is accompanied by evaporative water loss. The higher the temperature of the thorax, the more evaporation takes place, and the greater the amount of cooling. Evaporation then removes 8-9% of the total heat produced, but the difference in thoracic temperature resulting from evaporative cooling during flight in dry air and in air fully saturated with water vapour is relatively insignificant (Heinrich 1975). Reduction of thoracic temperature, through the transport of heat by blood from the flight muscles to the abdomen, also takes place when there is a relatively large temperature excess. Physiological limitations place great importance on behaviour that ameliorates thermal and water stress. The locust's relatively low metabolic rate in flight, together with nervous control of the spiracles, appears to make it well adapted to the desert.

5.2.3.2 Reptiles

Thermoregulatory panting is a common response of reptiles to overheating. It is not always a direct one, however: the interaction between physiological and environmental factors may be extremely complex. The threshold temperature for the onset of panting in *Amphibolurus muricatus*, for example, is affected by light intensity via the parietal organ and lateral eyes, and is elevated in light of high intensity (Firth and Heatwole 1976). Temperature regulation in land turtles has been reviewed by Sturbaum (1982). In addition to behavioural thermoregulation, he discussed evaporative cooling, cardiovascular responses, metabolic rates, ventilation and the role of the carapace in heating and cooling (Sect. 4.2.2.2). Copious salivation in response to heat occurs in young African desert tortoises *Geochelone* (= *Testudo*) *sulcata* (Fig. 12b), which can maintain a body temperature of about 40.5 °C at an ambient temperature of 55 °C (Fig. 46). The rate of transpiration in dry air takes place mainly through the lungs in baby animals, but cutaneous transpiration accounts for about half the evaporative water loss from older tortoises (Cloudsley-Thompson 1970a). In adult *Terrapene ornata*, both salivary frothing and urination take place when the cloacal temperature reaches 32.3-40.5 °C, and both assist in thermoregulation. Whereas heart rate is directly dependent upon body temperature, salivation may be triggered by both the temperature of the body and the brain. It is suggested that a function of the enlarged bladder of turtles may be to store urine for emergency thermoregulation (Riedesel et al. 1971). This species maintained core temperatures at least 8.5 °C below a 48 °C experimental ambient temperature for as long as 4 h, and 10.5 °C below a 51 °C experimental temperature for 3 h (Sturbaum and Riedesel 1974). In contrast, the thermoregulatory mechanism of *T. carolina* and other non-desert species breaks down in such an extremely hot environment (Sturbaum 1982). The results of heating whole animals and the brainstem of *T. ornata* to different temperatures indicate that there is an integration of central, peripheral and

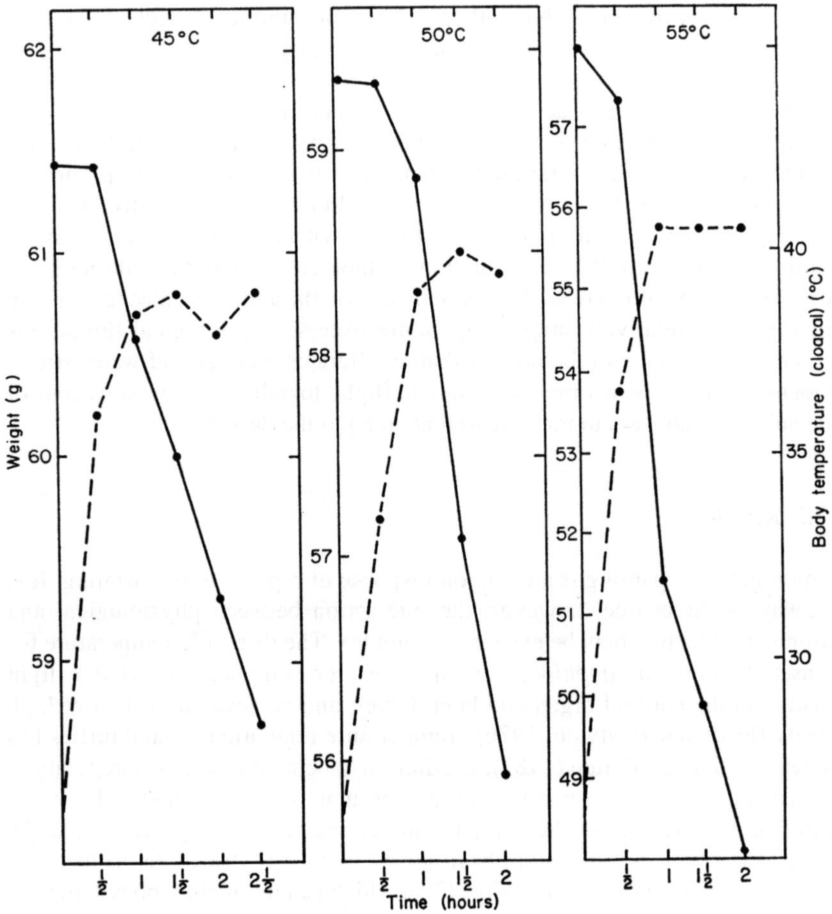

Fig. 46. Loss of weight and increase in body temperature of a baby *Geochelone sulcata* during consecutive exposures of 30 min in dry air at various high temperatures. Weight; —•— body temperature --•-- (Cloudsley-Thompson 1970a)

environmental sensory information that leads to salivation, urination and increased respiratory cooling (Sturbaum and Riedesel 1974). Thermoregulatory salivation has also been recorded in *Gopherus agassizii* by McGinnis and Voigt (1971) and in *Testudo graeca* by Cloudsley-Thompson (1974).

5.3 Excretory Water Loss

The employment of uric acid or guanine for nitrogenous excretion, originally associated with the evolution of cleidoic eggs and life on land (see Cloudsley-

Thompson 1988a) results in the pre-adaptation of insects, arachnids and reptiles for life in arid environments. Further desert adaptation, therefore, can only be a matter of degree. Nitrogenous excretion by insects has been reviewed in detail by Cochran (1985) and will not be discussed further here.

Uric acid and guanine are non-toxic and extremely insoluble, so they can be eliminated in a crystalline form without loss of water. This process contributes to the production of hypertonic urine which, in some desert arthropods, approaches the concentrations achieved by desert-adapted mammals. Similar processes produce dry faecal matter which, in insects, is eliminated with the urine as part of the excreta. The extent to which water is reabsorbed depends largely on external and internal moisture conditions. When partially dehydrated, desert species typically extract a higher percentage of water from their faeces (Hadley 1982).

Recent studies of water transport across the gut of the desert scorpion *Hadrurus arizonensis* illustrate the role of the ileum in water conservation (Ahearn and Hadley 1977). It was found that high levels of sodium ions in the lumen favour the transport of water from the gut into the haemolymph; while the uptake of water by the haemolymph is inhibited when the level of potassium ions is high. This mechanism appears to be adaptive, for it was also shown that concentrations of sodium surpass those of potassium in the ileum of dehydrated and starved scorpions. Consequently, water is conserved at times when its conservation is important. In freshly fed scorpions, on the other hand, excessive hydration is reduced by the inhibitary action of potassium ions on water transport into the haemolymph.

Economy in the use of water is, likewise, an abiding principle in the physiology of urinary excretion in reptiles. Like arthropods, they may be considered to be pre-adapted to desert conditions. The ability to minimize excretory loss, however, is associated with the lack of a loop of Henle in the reptilian nephron. Reptiles, therefore, cannot produce hypertonic urine, although *Amphibolurus maculosus*, a lizard which inhabits dry salt lakes in South Australia, voids a urine that is hyperosmatic to the plasma, the urine being concentrated in the cloaca (Braysher 1976) and not in the kidney as it is in birds and mammals. Renal function in reptiles has been reviewed by Dantzler (1976).

Most desert tortoises, lizards and snakes are either partially or completely uricotelic, although urate may not be excreted only as uric acid. When urinary pellets contain the electrolytes Na, K and NH_4, the low solubility of these cations, together with the high content of urate, low chloride content and high urine pH, suggest that they are bound as urate salts (Minnich 1972). Tortoises, in particular, show considerable temporal variations in the ratio of urea to uric acid excreted, and changes from predominantly uric acid to urea metabolism and vice versa have been observed within a single individual in several species. In an important review of the use of water by reptiles, Minnich (1982) pointed out that although these animals tolerate greater changes in the volumes of their body fluids and in the osmotic concentrations of their extracellular fluids than do birds and mammals, they regulate their water and electrolyte fluxes both physiologically and behaviourally; they are frequently selective in their choice of foods, avoiding those with high concentrations of electrolytes. When conditions are extreme, they may cease feeding, restrict activity and select favourable microclimates. Many species, too, are opportunistic drinkers, as discussed below (Sect. 5.6).

Water and mineral metabolism in Reptilia have been reviewed by Dantzler and Holmes (1974) who pointed out that lizards are the only terrestial reptiles which have been shown to possess extrarenal excretory organs: their salt glands secrete a fluid rich in potassium. In common with the marine iguana, desert reptiles are endowed with nasal glands which have the ability to adjust the relative amounts of sodium and potassium in the excretory fluid, depending upon the relative concentrations of these ions in the food or saline load (Templeton 1964b, 1966). Thus, in *Dipsosaurus dorsalis*, the rate of potassium excretion in the fluid from the salt gland increased up to four times following the injection of either sodium or potassium chloride. In the less xerophilic spiny lizard *Sceloporus cyanogenys*, the excretion of potassium increased in response to injections of either sodium or potassium chloride, and no adjustment in the Na^+: K^+ rate was apparent in the fluid. There appeared to be no extrarenal mechanism capable of responding to high loads of sodium. Lemire et al. (1980) and Lemire and Vernet (1982, 1983) obtained comparable results with *Uromastix acanthinurus*. They found that the nasal glands were the primary source of excretion of chloride ions, but that the $Cl^-/(Na^+ + K^+)$ ratio may vary. This variation allows the lizards to adapt to natural variations — seasonal or geographic — in their plant food. Salt-excretory nasal glands clearly play an important role in the elimination of the salt absorbed in the diet of plant-eating desert reptiles.

5.4 Osmoregulation

Despite their efficient physiological mechanisms for avoiding water loss, even desert-adapted arthropods and reptiles sometimes experience, and are capable of surviving, severe desiccation without the osmotic pressure and composition of their haemolymph changing to such an extent as to adversely affect the functioning of the body.

5.4.1 Arthropods

Many desert arthropods can survive losses of body water in excess of 30% of their original wet weight (Edney 1974, 1977; Hadley 1974; etc.). Despite considerable reduction in the volume of the haemolymph, osmotic pressure is regulated during dehydration. Edney (1968) measured changes in the volume of the blood, its water content and osmotic pressure during dehydration in the North American desert cockroach *Arenivaga investigata*. He found that water was lost proportionally from the blood and from the rest of the body, so that a decrease in the volume did not cause as much rise in the osmotic pressure as might be expected. This mechanism has subsequently been disclosed in parallel studies on other insects.

During desiccation, the spirostreptid millipede *Orthoporus ornatus* displays both osmotic regulation and tolerance patterns, depending on sex and on the duration of desiccation (Riddle et al. 1976). Crawford et al. (1986) investigated seasonal water balance in *Archispirostreptus syriacus* (Spirostreptidae) from mesic and xeric Mediterranean environments. Seasonal levels of body water were similar among all populations, but most variable in those from xeric habitats. In all populations, water increased in the cuticle and tissues during summer, while it decreased in the gut and haemolymph. The osmolarity of the haemolymph also increased in summer, mainly from xeric populations. Moderate dehydration seemed to cause shifts of water from the gut to other compartments of the body. Osmotic regulation appears therefore to be less efficient among millipedes than other arthropods.

In marked contrast are tenebrionid beetles. Specimens of *Trachyderma philistina* can lose up to 35% or more of their body weight, without ill effects, whilst kept under dry conditions. When supplied with water after such dehydration, they drink rapidly (Sect. 5.6.1) and quickly surpass their original weight. Even so, only part of this water passes from the gut to the haemolymph and thence to the tissues during the first day after drinking. Changes in the osmotic pressure of the haemolymph, during both dehydration and subsequent rehydration, are slight. Organic molecules, rather than inorganic ions, play the principal role of osmolar effectors (Broza et al. 1976).

According to Riddle et al. (1976), *Eleodes hispilabris* controls haemolymph osmoregulation fairly precisely, as does *Stips stali*, a nocturnal tenebrionid from the Namib (Naidu and Hattingh 1986) and the sympatric *Onymacris unguicularis*. Active beetles, however, must drink condensed fog (Sect. 5.6.1) in order to maintain water balance, while inactive beetles can maintain water balance either by eating seeds, or by simply metabolizing fat. Little change is observed in ratios of haemolymph and total body water to dry mass when fogs occur, but there are significant changes in the osmotic pressure of the haemolymph (Cooper 1982). The congeneric *O. plana* has impressive osmoregulatory abilities (Nicolson 1980), so the possession of a diuretic hormone is hard to explain (Nicolson and Hanrahan 1986). Perhaps the reabsorptive processes in the cryptonephridial system have the capacity to recover all the water produced by the Malpighian tubules. If so, the action of the diuretic hormone may be, not to remove water, but to promote more rapid removal of metabolic wastes from the haemolymph. The production of these wastes is naturally accelerated by high temperature (Cloudsley-Thompson 1962c; Ahearn 1970a).

Unlike beetles, some scorpions, such as *Paruroctonus aquilonalis* of New Mexico, seem to tolerate increasing osmolarity during desiccation (Riddle et al. 1976). No osmotic regulation was observed during dehydration of the East African *Lychas burdoi* (Buthidae) from semi-arid regions of East Africa (Bjerke et al. 1989). Although the South African desert buthid *Parabuthus villosus* (Fig. 47) has very low cuticular permeability, it does not inhabit burrows and is occasionally active during the day on the hot surface of gravel plains in the Namib desert. Yet, analysis of its haemolymph during prolonged desiccation, nevertheless shows good osmotic and ionic regulation, in marked contrast to the mesic

Fig. 47. *Parabuthus villosus* (Buthidae), Namib desert

burrowing scorpionid *Opisthophthalmus capensis* from the region of Table Mountain (Robertson et al. 1982). So it seems that desert scorpions may be better able to osmoregulate than species from less arid regions.

5.4.2 Reptiles

Like desert arthropods, desert reptiles can withstand a considerable degree of dehydration without ill effects: e.g. a loss of over 50% of its initial body weight in the case of the nocturnal Sudanese gecko *Tarentola annularis* (Cloudsley-Thompson 1965a), *Dipsosaurus dorsalis, Uma scoparia Cnemidophorus tigris, Callisaurus draconoides* and even in that of the mesic *Sceloporus occidentalis* (Munsey 1972). Moreover, many desert reptiles have considerable powers of osmoregulation. For example, when *Sauromalus obesus* becomes dehydrated, the fractional fluid volumes and plasma electrolytes are maintained at normal levels (Nagy 1972).

For many years, the idea that water could be absorbed from the reptile bladder or cloaca was discounted but, more recently, evidence has accumulated suggesting that, in tortoises at any rate, uric acid and urates are deposited in the bladder, and water is reabsorbed (Dantzler 1976; Minnich 1979, 1982). This allows the animals to withstand large water losses, yet maintain stable osmotic concentrations in the intracellular and extracellular fluid compartments. Although such a fluid reserve would appear to be a useful adaptation for reptiles in arid environments, most desert species lack a urinary bladder. Although present in tortoises and turtles, the bladder is absent from snakes, and many species of lizards also lack one when adult.

Neonatal lizards of the viviparous species *Sceloporus jarrovi* possess, at birth, a urinary bladder which contains a large amount (14% of body mass) of dilute urine. After birth, no additional fluid is added to the bladder. On the contrary, the fluid it contains declines in volume and increases in osmotic concentration as the bladder degenerates to the vestigial organ found in adults. To the young, however, it is a useful extrarenal osmoregulatory organ that can buffer body water compartments against osmotic perturbation during growth (Beuchat et al. 1986).

In an interesting comparison of seasonal patterns of water and energy balance among desert vertebrates of the Mojave desert, Nagy (1988) found that the two herbivorous reptiles, chuckwalla lizards (*Sauromalus obesus*; Fig. 15a), and desert tortoises (*Gopherus agassizii*; Fig. 12a), both husband their water and energy budgets remarkably well, but differently. The chuckwallas maintain precise regulation of the osmotic pressure and fluid volume of the blood. This restricts them to feeding only on green, moist vegetation in spring. Tortoises, on the other hand, tolerate wide swings in their osmotic and fluid balance. Consequently, they are able to eat dry vegetation and drink rainwater during summer and autumn. This flexibility may account, in part, for the observation that desert tortoises are able to lay eggs annually, even during drought years (Turner et al. 1986), whereas chuckwallas cannot.

Osmoregulation in arthropods has been reviewed by Edney (1977), in reptiles by Bentley (1976), to which authors the reader is referred for details of the physiological mechanisms involved.

5.5 Behavioural Regulation of Water Loss

Whereas thermoregulation in arachnids and reptiles is achieved largely by behavioural means, water relations are almost entirely under physiological control. There are, however, a few ways in which water loss can be reduced behaviourally. For instance, the maintenance of a constant rate of oxygen consumption over a given period of time results in greater loss of water than would the employment of a cyclic breathing pattern. This subject has not been investigated to any great extent, but it is known to occur in carabid and tenebrionid beetles (Ahearn 1970c;

Bartholomew et al. 1985; Louw et al. 1986), scorpions (Hadley and Hill 1969), and tortoises (Cloudsley-Thompson 1974).

In thermal gradients, desiccated lizards select significantly lower body temperatures than do hydrated specimens. They also show a decrease in activity, tending to remain burrowed in the substrate of a thermal gradient. Reduction in body temperature and activity presumably functions to decrease water loss (Crowley 1987).

Brooding is a rare phenomenon in lizard behaviour except in the genus *Eumeces* (Scincidae) where it is found in all ovigerous species. Although several different functions have been attributed to this trait, it does not appear to be thermoregulatory as in pythons (Sect. 4.4.1). The only controlled laboratory experiments have been carried out on *E. septentrionalis*, a semi-fossorial species that inhabits the semi-arid prairies of North America. According to Somma and Fawcett (1989), brooding by female prairie skinks reduces egg mortality. It regulates the amount of water exchanged across the eggshells by altering the proportion of the shells that is in contact with the substrate, the atmosphere, the female's body and surrounding eggs, and by altering the volume of the nest cavity.

5.6 Water Uptake

Although, at one time, scorpions and many other desert arthropods were believed never to drink liquid water, this is now known not to be the case. When dehydrated, some scorpions drink readily (Hadley 1974) (first recorded by the present writer in 1958) as do woodlice, Solifugae, spiders, myriapods and insects (Edney 1977). The rate and quantity of water drunk varies greatly; this must to some extent depend upon the amount absorbed with the food. The solifugid *Galeodes granti* scarcely drinks at all in nature although it can do so (Cloudsley-Thompson 1961c,d) but devours large numbers of insects and other arthropods, obtaining moisture from the haemolymph of its prey, as do scorpions and other predatory arthropods. Some arthropods probably eat more in low than in high humidities, or eat more moist than dry food when they are short of water. *Onymacris* spp. certainly do this (Edney 1977). Both metabolic and pre-formed water are obtained from food. Whether production of the former equals or exceeds the water lost in metabolism depends upon the efficiency of excretion and the control of water loss during respiration.

Drinking has not been observed in all reptiles, but takes place in many when the opportunity presents itself (Minnich 1979, 1982; Sect. 5.6.1). According to Nagy (1972), *Sauromalus obesus* does not drink, and *Tarentola annularis* drinks very little, even when severely desiccated (Cloudsley-Thompson 1965a). On the other hand, the Californian legless lizard *Anniella pulchra* is able to imbibe soil water through its mouth (Fusari 1985) as can amphisbaenians. The unique uptake of water by *Moloch horridus* is described below (Sect. 5.6.1).

5.6.1 Drinking and 'Fog Basking'

The coastal desert of Peru and northern Chile, like the Namib, is almost rainless except in exceptional years of *el Nino* (= Christ Child), when the cold Humboldt current is replaced by a warm equatorial current from the Gulf of Guayaquil, and extremely heavy rainfalls. Such unpredictable rain is far less important to the flora and fauna of the desert than is the fog or *garua*. This forms above the cold ocean current and is blown inland by the southwesterly winds that arise as a result of the warming of the land. Wetting fogs occur near the coast mainly during the winter months for, in summer, the land heats so rapidly that there is little condensation. They make possible the development of scattered *loma* (= hill) vegetation where condensation takes place as the fog is carried upwards across the slopes of the cordilleras. A characteristic fauna, associated with *loma* vegetation consists of woodlice, flies, carabid beetles, the black widow spider *Latrodectus mactans* and the scorpion *Hadruroides lunatus* (Walter and Breckle 1986). The most abundant groups in the *lomas* of southern Peru, according to Péfaur (1981), are Acari, Collembola, Homoptera, Coleoptera, Diptera and Formicidae. These show a sudden numerical increase in September, prior to the emergence of new plants, indicating a reliance on perennials for food and shelter rather than on less predictable annuals.

Drinking droplets of moisture from fog condensed on the body is an important source of water uptake for both arthropods and reptiles, not only in the Atacama and Peruvian deserts, but in the Namib, where it was first observed, the Negev desert and the Wahiba Sands, Oman. Louw (1972) was the first to describe the role of advective fog in the water economy of Namib desert arthropods and reptiles. The psammophilous tenebrionid beetle *Lepidochora discoidalis* (= *argentogrisea*) and the dune lizard *Aporosaura anchietae* have both evolved a physiological dependence upon fog water, and are capable of a considerable degree of water storage. The insectivorous *Meroles cuneirostris* (Fig. 48) does not drink dew.

L. argentogriseus can obtain sufficient nutrients from dry vegetation to satisfy its calorific requirements but also requires a source of free water. This is supplied by fog which condenses on the elytra, and is subsequently drunk and stored in the soft tissues. *A. anchietae* and the side-winding viper *Bitis peringueyi* likewise drink droplets of water from fog that condenses on their bodies. *A. anchietae*, in particular, stores a considerable quantity — up to 10% of its body weight — at first in its stomach; afterwards, this is transferred to the caecum.

'Fog basking' by *Onymacris unguicularis* was first observed by Hamilton and Seely (1976). Beetles of this species are normally day-active, preferring the warm part of the day, and remain buried in the sand of dune slip faces at night. When there is fog at night, however, they emerge and climb the slip face to the crest. Here, they adopt a characteristic head-down stance (Fig. 49), facing into the fog-laden wind while water trickles down the body to the mouth. Although *Physadesmia globosa* is not a fog-basking species, it also emerges on foggy nights and sucks moisture from damp sand (Sect. 5.6.2). The posture that it adopts,

Fig. 48. *Meroles cuneirostris* (Lacertidae) Namib desert. This individual has a forked tail, probably an abnormality of regeneration

Fig. 49. Fog basking by *Onymacris unguicularis* (Tenebrionidae), Namib desert

whilst drinking, is similar to the head-down, fog-basking stance of *O. unguicularis* and may provide an explanation to the evolution of the adaptive behaviour of the latter (Cloudsley-Thompson 1990).

Several genera and species of tenebrionid beetles make use of fog, dew or rain in hyper-arid environments by constructing on the dune surface small mounds of sand, and trenches which contain enough free water for active intake during the early hours of the morning. Extraction of water ceases at sunrise when the mounds and trenches begin to disintegrate. Three species of *Lepidochora* are known to do this in the Namib desert. Trenches are constructed perpendicular to fog winds and concentrate moisture during fogs. The beetles return along the ridges of the trenches which they have excavated, extracting water from them. The water content of a population of these beetles increased by 13.9% during one fog (Seely and Hamilton 1976). Further examples of the responses to fog of Namib dune beetles are given by Seely (1978, 1979) and Seely et al. (1983).

Dependence upon moisture from advective fog has been recorded in various desert arthropods, including isopods, bugs, ants, carabid, coccinellid and tenebrionid beetles of the Negev desert (Broza 1979). There are, on average, 42 foggy nights per year in the northern part of the desert area of Israel, and fog there plays approximately the same role as dew in the ecosystem. In the Wahiba Sands of Oman, fog and dew are drunk by *Ammogiton* and *Erodius* spp. Mounds, trenches and subsurface tunnels in moist sand are constructed in a manner characteristic of each species (Büttiker and Anderson 1988).

Larvae of tiger beetles (Cicindelidae) are sit-and-wait predators that occupy permanent burrows in the ground. They wait at the tops of these with their heads flush with the surface substrate. When a suitable arthropod approaches within striking distance, the larva springs backward out of its burrow, seizes the prey with its mandibles and pulls it back into the burrow. Water balance is maintained by ingesting moist sand whilst inside the burrow, and absorbing the water in the digestive tract (Hadley et al. 1990). According to Broza (1979), many desert insects in Israel take water from damp soil and hyposcopic vegetable matter on foggy nights.

Although desert scorpions (*Hadrurus arizonensis*) are apparently unable to take up water from near-saturated atmospheres or moist substrates, regardless of their state of hydration (Hadley 1970b), the semi-montane *Diplocentrus spitzeri*, especially in its second instar, seems to be able to regain moisture from moist soil when dehydrated without ingesting soil particles (Crawford and Wooten 1973). The latter species loses water by transpiration at a rate of 8.49 $\mu g\ cm^{-2} h^{-1} mmHg^{-1}$ (cf. 0.98 from *Buthotus minax*) and is therefore not particularly well adapted for life in a xeric microclimate (Sect. 5.2.1.1). *Mastigoproctus giganteus* likewise absorbs moisture from damp sand (Ahearn 1970b; Crawford and Cloudsley-Thompson 1971), while spiders, too, have long been known to be able to take up water from moist soil. In the case of the Australian wolf spider *Geolycosa godeffroyi*, water cannot be extracted from near-saturated air but is available from the soil when its water content exceeds as little as 11% (Humphreys 1975).

Desert tortoises, *Gopherus agassizii* (Fig. 12a), often become active during and after the infrequent showers and thunderstorms they experience. They drink from temporary pools, even when air temperatures are suboptimal (9-15 °C) and sometimes construct shallow catchment basins which hold water for several hours (Medica et al. 1980). *Geochelone sulcata,* of the African Sahel savanna goes further, actually digging depressions in the soil at the time of rain, from which it drinks (Grubb 1971).

The water-collecting mechanism of *Moloch horridus* has been re-examined by Gans et al. (1982) who showed that the collection and transport of water are more complex than previously realized, and occur in different regions of the integument than previously thought to be the case (Bentley and Blumer 1962).

The animal's spines, long assumed to represent a defensive device, serve also to attract molecules of water vapour which condense upon them and are conveyed by capillarity to the mouth (Gans et al. 1982). Its ecological equivalent, the iguanid *Phrynosoma m'calli*, exhibits many traits similar to those of the agamid *M. horridus*, but does not appear to possess the ability to utilize its skin for the acquisition of water (Mayhew and Wright 1971). The Californian legless anguid lizard *Anniella pulchra*, however, is able to absorb water if its mouth is in contact with sand having a moisture content above 6.9% (Fusari 1985).

Kaufmann and Pough (1982) found that the efficiency of assimilation of pre-formed water increased 10% between 34 and 40 °C in *Dipsosaurus dorsalis*, according to the digestibility of the food. In summer, desert iguanas spend 90% of the daylight hours in their burrows where the assimilation of pre-formed water equals the increase in evaporative water loss as temperatures increase.

As already mentioned, *Bitis peringueyi* drinks drops of water from fog that condenses on its body (Louw 1972). The viper flattens its body dorsoventrally thereby increasing the surface upon which moisture can collect. The response occurs in hatchlings which suck moisture which collects in a crease formed by pressing the rostrum against the lateral body surface (Robinson and Hughes 1978).

5.6.2 Uptake of Soil Moisture

A source of water not neglected by desert arthropods is soil moisture, and members of numerous taxa have been shown to ingest moist soil and extract the water in their guts, or else to drink capillary water. Desert millipedes, such as *Orthoporus ornatus* which takes up moisture from the substrate both orally and anally (Crawford 1972) and *Harpagophora nigra* provide examples of the latter (Crawford and McClain 1983).

5.6.3 Absorption of Atmospheric Moisture

The uptake of water vapour from unsaturated air has been the subject of much investigation and numerous speculations. Among desert arthropods, it takes place in adult females of *Arenivaga investigata*, where it was first described by Edney (1966b), in *Ctenolepisma terebrans*, a sand-dwelling thysanuran of the Namib desert (Edney 1971b) and in the larvae of the tenebrionid beetles *Onymacris plana* and *O. marginipennis* (Coutchié and Crowe 1979). These insects are incapable of flight, a characteristic of all arthropods that are able to absorb moisture from unsaturated air (Edney and Nagy 1976; Edney 1977). No reptile can absorb atmospheric moisture, so there are no parallels between arthropods and reptiles in this context.

C. terebrans, when dehydrated, gains weight at relative humidities of 47.5% or above, until its normal water balance has been restored (Edney 1971b), while critical equilibrium humidities of both ixodid and argasid ticks (Acari: Metastigmata), which also possess this unusual ability, vary between 70-90% RH (Knülle and Rudolph 1983).

Absorption of atmospheric moisture has been studied in greater detail in *Arenivaga investigata* than in any other desert species. Edney (1966b) found that females and nymphs can take up water at humidities above 82.5% RH between 10 and 30 °C. The osmotic pressure of the haemolymph is well regulated against the effects of dehydration and rehydration. Abrasion of the cuticle causes increased transpiration, but the damage is soon repaired. The physiology of water absorption in *A. investigata* has been studied in considerable detail by O'Donnell (1977, 1981, 1982) who showed that buccal structures are involved in relative humidities above 81%; vapour condenses onto a fluid layer covering two protrusible bladders which are lateral diverticula of the hypopharynx. The fluid is held in the interstices of fine cuticular hairs which densely cover the bladder surface (O'Donnel 1977, 1981). Absorption by *A. investigata* differs markedly from the solute-dependent schemes that have been proposed for other arthropods (see O'Donnell 1982; Hadley 1984).

5.7 Conclusions

The problem of dehydration in desert animals reflects both the increased saturation deficiency of the atmosphere and, sometimes in the case of reptiles and large insects whilst in flight, the employment of evaporative cooling. Water is lost, both as liquid and as vapour, through a variety of paths, depending upon the species concerned and its state of hydration. Although desert arthropods and reptiles to some extent avoid water loss by behavioural means, water conservation, unlike thermoregulation, is primarily physiological. The major avenues of evaporative water loss are through integumental and respiratory transpiration and excretion; of water uptake by drinking or absorbing moisture from the soil and, in a few specialized arthropods, from unsaturated air.

6 Seasonal Activity and Phenology

6.1 Introduction

Rainfall is the major environmental factor affecting biological productivity in the desert biome (Sect. 2.2) and, consequently, the food available to arthropods (Cloudsley-Thompson and Idris 1964; Crawford 1981; Louw and Seely 1982; Wallwork 1982; Whitford 1986; etc.) and reptiles (Mayhew 1966a,b, 1967, 1968; Ballinger 1977; Dunham 1978; Vitt et al. 1978; Worthington 1982).

Seasonal and yearly fluctuations in Chihuahuan desert lizard communities, for example, are related directly to the relative abundance of arthropods which, in turn, depends upon rainfall. The density of adult teiids, *Cnemidophorus tigris* (Fig. 9a), nearly doubled after a relatively wet period of 2 years, and this was attributed to the survival of more hatchlings than usual during the winter. In contrast, the numbers of *Sceloporus magister* remained constant, probably because this species is more terrestial in habit (Whitford and Creusere 1977). The influence of seasonal changes on the daily cycles of activity of desert arthropods and reptiles has already been discussed (Sect. 3.4). In the present chapter, attention will be focussed upon seasonal avoidance of climatic extremes, diapause, reproductive cycles and the like.

6.2 The Arthropods of Temporary Rain Pools

Many desert arthropods survive in the desert by enduring long periods of unfavourable climatic extremes in a state of arrested development or diapause. During diapause, metabolic demands are minimal, water loss is extremely low and resistance to high temperatures is marked. Natural depressions on desert plains, known as *playas* (Heathcote 1983), or *panne* in the Kalahari, remain dry and apparently lifeless for years. Within a day or two after being flooded with rain, however, bacteria, algae and protozoans, which had previously been encysted, appear in large numbers. The diapause eggs of phyllopod crustaceous Anostraca, Notostraca and Conchostraca, hatch almost immediately and develop very rapidly.

Intense insolation in the desert environment causes rapid evaporation, high temperatures and, consequently, a low O_2 tension in the water. Under these unstable conditions, the animals have to grow swiftly and, if they are to survive,

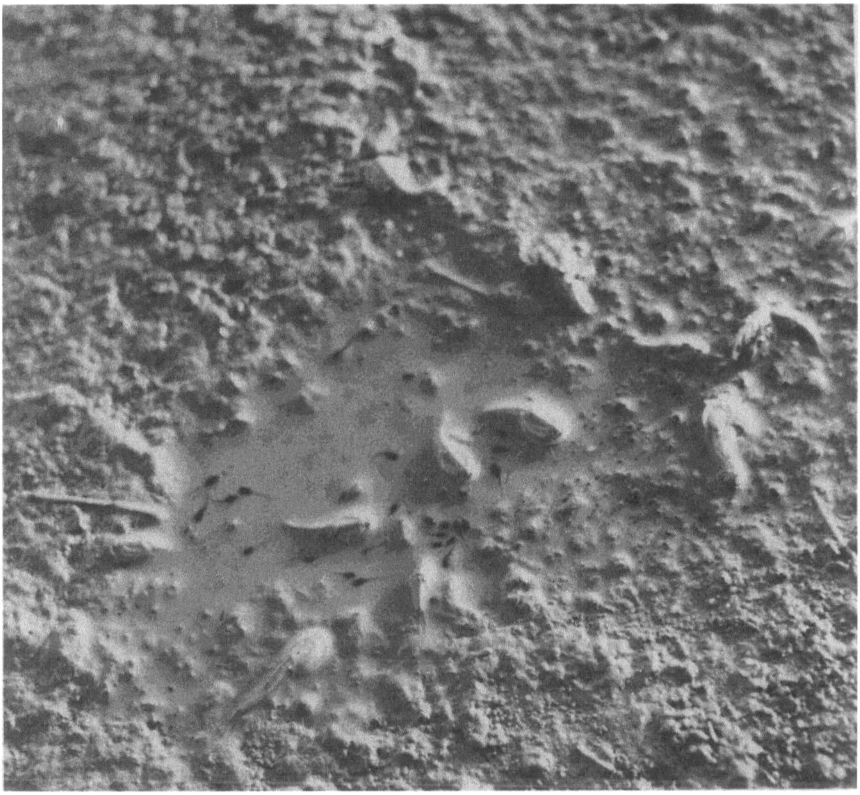

Fig. 50. Drying desert pool in Sudan with stranded *Triops granarius* and tadpoles

reproduce before the pools dry up (Fig. 50; Louw and Seely 1982). A considerable amount of biological information is given by Rzóska (1961, 1984). *Triops* spp. appear in pools around Khartoum after 7 days: in Europe, after 2-3 weeks. The life span at Khartoum is limited to 20 days: in Europe, to 2-3 months. Conchostraca appear at Khartoum within 3 days and are mature after 5. In Europe, they appear after 20 days and live for many weeks. Ephemeral life in temporary pools is not restricted to deserts, but there the periods of dormancy are longer and more unpredictable than elsewhere.

In their review of the adaptational biology of the inhabitants of temporary ponds, Belk and Cole (1975) distinguished between obligatory diapause, which occurs every generation and is endogenously controlled, and facultative diapause, which appears to be evoked by environmental factors that predict a change to unfavourable conditions. From an ecological viewpoint, the former would be the more appropriate in predictably cyclic environments, the latter in less certain situations. Only by experimental study, however, can the situation be determined for any particular species. Although Belk and Cole (1975) cited several examples of facultative diapause among the phyllopods of desert pools, Louw and Seely

(1982) considered them to have drawn a distinction between the two that was perhaps too sharply defined, and that the life histories of these animals are probably controlled by the interaction between an endogenous rhythm and exogenous cues.

The egg is the most usual stage in which diapause occurs among desert Crustacea. The metabolism of arthropod eggs is so low during diapause that their survival is possible for many years. Even so, not all the inhabitants of desert pools undergo diapause as eggs. Rzóska (1961) found that *Metacyclops minutus* (Copepoda) can survive extremely hot conditions in the Sudan, in the resting copepod stage of development. Eggs of *Triops granarius* may survive 16 h at 98 °C (Carlisle 1968), but the lethal temperature of the adults if 40 °C for 2 h or 34 °C for 24 h (Cloudsley-Thompson 1965b); while the lethal temperature of *Streptocephalus* spp. is 35 °C for 24 h (Cloudsley-Thompson 1966).

The significance of diapause lies in the fact that not every egg hatches the first time that it is wetted. Some may not hatch until they have been moistened on several occasions. In consequence, a light shower will not stimulate the hatching of every egg when there is insufficient water in the pools to last for the entire period of development. Only some individuals hatch and die after each light shower.

Not all the inhabitants of temporary desert rain pools withstand dry periods in a state of diapause. The eggs and developmental stages of some arthropods can withstand total dehydration in a state of cryptobiosis. A classical example of this is provided by the larvae of the chironomid midge *Polypedilum vanderplanki* which inhabits shallow rock pools in Nigeria and Uganda. During the rainy season, these pools may fill and dry several times. Each time that rain falls the larvae absorb water through their integument and begin feeding within an hour. In the dry state, they can withstand exposure to 106 °C for 3 h and 200 °C for 5 min. They also tolerate immersion in liquid air (-190 °C) or liquid helium (-270 °C) yet subsequently produce normal adults. For the hydrated larvae, on the other hand, thermal extremes are fatal (Hinton 1960). Certain Australian chironomid larvae may also withstand total dehydration; further examples are cited by Edney (1977).

Chironomus imicola of tropical Africa has no such ability. If it is to survive, it has to leave the water as a winged adult before the pool dries. Selective pressure on its life cycle, however, fits it to the ephemeral nature of its habitat. The larvae respond to population density in the rain pools throughout their development. When densities are low, safe early emergence produces small females. As the pools evaporate, the density of larvae increases and larger, but fitter, adults emerge. Finally, at very high density, when extinction of the habitat is imminent, females emerge earlier, but are again larger. The existence of females, both large and early, is surprising; but they endure a high risk, only about 2% of the larvae that produce them reaching adulthood (McLachlan 1983). Belk and Cole (1975) outlined the ways in which the inhabitants of temporary ponds, whilst in the hydrated state, are able to cope with adverse conditions including thermal extremes, low oxygen concentration and high salinity in the surrounding water. To all of these exigencies, the larvae become rapidly adjusted. Behavioural as well as physiological adaptations are found. For example, phyllopods of the Sonoran desert swim to the surface of oxygen-depleted water and stir it with their appendages (Hillyard and Vinegar 1972). *T. granarius* shows positive rheotaxis and swims against currents: this tends to prevent it from being stranded by strong winds on the leeward sides of pools (Cloudsley-Thompson 1966). Adaptations to abiotic stress conditions in temporary desert pools are discussed by Crawford (1981) as well as by Belk and Cole (1975).

6.3 Flooding

Paradoxically, one of the hazards that desert arthropods have to face, if they live in burrows, is flooding of the ground. A sudden thunderstorm can cause a deluge with widespread sheet flooding. This may explain why so many species of desert ants surround the entrances to their nest with walls of sand (Figs. 13a,b) which help to keep water out (Sect. 2.4.3). Trapdoor spiders, too, have countered the risk of flooding with a variety of burrow structures and impregnable entrances. Examples from the Australian desert regions are described in detail by Main (1976, 1982) and therefore will not be discussed further here.

6.4 Diapause, Aestivation and Seasonal Patterns of Activity

Seasonal changes of the environment in hot, arid regions result from one or other of two factors, which may or may not coincide. These are seasonal precipitation and seasonal periods of cooler weather even if rain does not fall. Many desert organisms avoid climatic extremes by seasonal as well as by diurnal rhythms of activity. These are mutually interactive because diurnal locomotory rhythms often show seasonal influences (Sect. 3.4). Such changes are reflected in the numbers and stages of development of the fauna.

6.4.1 Arthropods

Dry season diapause is very common in desert insects, but is triggered neither by lack of rain nor by a decrease in relative humidity (Denlinger 1986). Adults of *Adesmia bicarinata* (Tenebrionidae) begin to appear in small numbers in Egypt during late October. Their population then gradually increases, reaching a peak in March. By the end of May, the beetles have disappeared completely, and only dead bodies can be found. During the hot season, the life cycle is continued by the larval and pupal stages of development. In the case of life cycles such as this, which are extremely common among desert insects and arachnids, diapause and quiescence may or may not occur in the developmental stages (see Cloudsley-Thompson 1965c; Abushama 1984). Either way, a vernal or rain fauna appears at the time of inflorescence when the desert is transformed by an abundance of plant and animal life. Flowers are visited by Lepidoptera, Coleoptera, Syrphidae, Bombyliidae and other Diptera. Termites extend their subterranean galleries to the soil surface and indulge in nuptual flights, while predators such as scorpions, Solifugae, ant lions (Myrmeleontidae), Heteroptera, wasps, Asilidae and predatory beetles glut themselves on an abundance of food.

With the rains, too, come the locusts. Desert locusts feed on the ephemeral grasses that spring up after rain, and later deposit their eggs (see Cloudsley-Thompson 1977e). Sexual maturation in *Schistocerca gregaria* is stimulated by the presence of terpenoids and other aromatic chemicals in *Commiphora myrrhae* and various other desert shrubs in the season of the monsoon-type rains. Carlisle et al. (1965) suggested that bud burst in these shrubs provides the environmental cue by which desert locusts gear their breeding seasons to the rains.

Desert arthropods often show seasonal acclimation in their thermal relations. The woodlice *Porcellionides* (= *Metoponorthus*) *pruinosus* and *Periscyphis jannonei* in the Sudan show rates of transpiration in dry air that are related to the xeric nature of their environment, but rapid acclimation in the rate of water loss does not occur. Differences between populations are probably the result of genetic adaptation. Both species, however, show quick acclimation to high and low temperature, not only regarding their time of survival in saturated air, but also their metabolic rates (Cloudsley-Thompson 1969b). In contrast, the North American desert millipede *Orthoporus ornatus* has less body water in autumn and winter than in spring and summer (Crawford 1978), as does *Archispirostreptus syriacus* in Israel (Crawford and Warburg 1982).

The ability of *Scolopendra polymorpha* to recover from supercooling (Sect. 4.3.3) suggests that this capability may be a factor in the overwintering physiology of the species (Crawford et al. 1975). Nevertheless, *S. polymorpha* is only moderately frost-resistant, which would account for the mortality that occurs during occasional periods of cold weather.

Although woodlice do not show seasonal acclimatory effects on the rate of transpiration (Cloudsley-Thompson 1977a), insects certainly do. As already mentioned (Sect. 5.2.1.1), Hadley (1977) found that the cuticle of *Eleodes armata* in summer, or in winter after acclimation to 35 °C, exhibits larger quantities of hydrocarbons and a higher percentage of long-chain components than do beetles in winter, or appropriate controls. These changes may well contribute to increased impermeability of the integument, although attempts to correlate the thickness of cuticular wax deposits with transpiration rates have generally been unsuccessful. On the other hand, increasing the number of carbon atoms in a hydrocarbon chain decreases the permeability of a membrane. Cuticular permeabilities of *Hadrurus arizonensis* and *Uroctonus apacheanus* (Vejovidae) likewise reflect the thermal and humidity regimes of their environments (Toolson and Hadley 1977; Sect. 5.2.1.1).

In a study of daily and seasonal activity among ground-living Tenebrionidae in the southern Namib and Kalahari ecosystems, S. Louw (1983) not only distinguished between the faunas of different microhabitats, but showed that the density and diversity of species reaches two peaks during the annual cycle. These peaks are both during the 2 to 3 months that succeed the rainy season. The first appears during the post-rain or autumn period, the second in winter. Throughout the remaining months of the year, diversity and density are relatively low. At the same time, during all seasons, the diversity of nocturnal species is higher than that of diurnal species. (Diurnal and seasonal periodicity in Namib desert tenebrionids is discussed by Wharton and Seely 1982.) The activity of Namib desert ants, like that of beetles, increases significantly after rain, with an outburst of foraging and nest construction (Marsh 1988). The brief activity periods of mydas flies (Mydidae) is also an adaptation to the extreme environment of the central Namib (Wharton 1982), where precipitation and strong winds serve as cues for emergence (Louw and Hamilton 1972; Holm and Edney 1973).

Although *Paruruoctonus mesaensis* (Vejovidae) is active on the surface of the Californian desert throughout the year, densities are highest in spring and autumn, lowest in mid-winter and mid-summer (Fig. 51; Polis 1980a). Over 5 consecutive years, the first appearance of newborn young was found to be synchronous throughout the population, and occurred in early August. Maturity is achieved in 19-24 months and growth is determinant. Three age groups can therefore be identified in the field: young animals less than 1 year of age, intermediate animals 1-2 years old, and adults more than 2 years of age. Since the females can live for at least 5 years, the adult age group includes animals from differently aged cohorts.

The surface density of adults and members of five cohorts were recorded from August 1973 through September 1976 in a 1400 m^2 quadrat (Fig. 51; Polis 1980a). The density of adults was highly correlated with insect abundance, but not that of the two younger age groups. Moreover, it was high during the mating season, and low during the mid-summer period when birth and brooding of the newborn young occurred. It appears, therefore, that adult activity is an evolutionary response to periods of high prey density, whereas the activity of the two younger age groups reflects long-term avoidance of adults and/or competition for food (Polis 1980a).

Giant velvet mites *Dinothrombium tinctorium* (Fig. 16) appear in the Sahara desert after rain. They are day-active while the sand is moist, but become crepuscular as it dries out. They have a high rate of water loss by transpiration, suggesting that they are not particularly well adapted to drought; are positively phototactic and negatively geotactic in dry sand, but dig burrows where it is damp (Cloudsley-Thompson 1962b). Their ecological equivalent in southern California is *D. pandorae* which likewise spends most of its adult life alone in a simple, vertical burrow, only coming to the surface after rain. Emergence is sometimes correlated with the swarming of termites upon which the adult mites feed exclusively. During spring, the mites remain some 10 cm below the surface, at the upper end of their burrows but, in summer, they descend to a depth of about 170 cm. Cold-induced dormancy occurs below 10-15 °C. During winter, therefore, the mites are dormant at night and become active only on sunny days. The larvae are parasitic on grasshoppers (Tevis and Newell 1962).

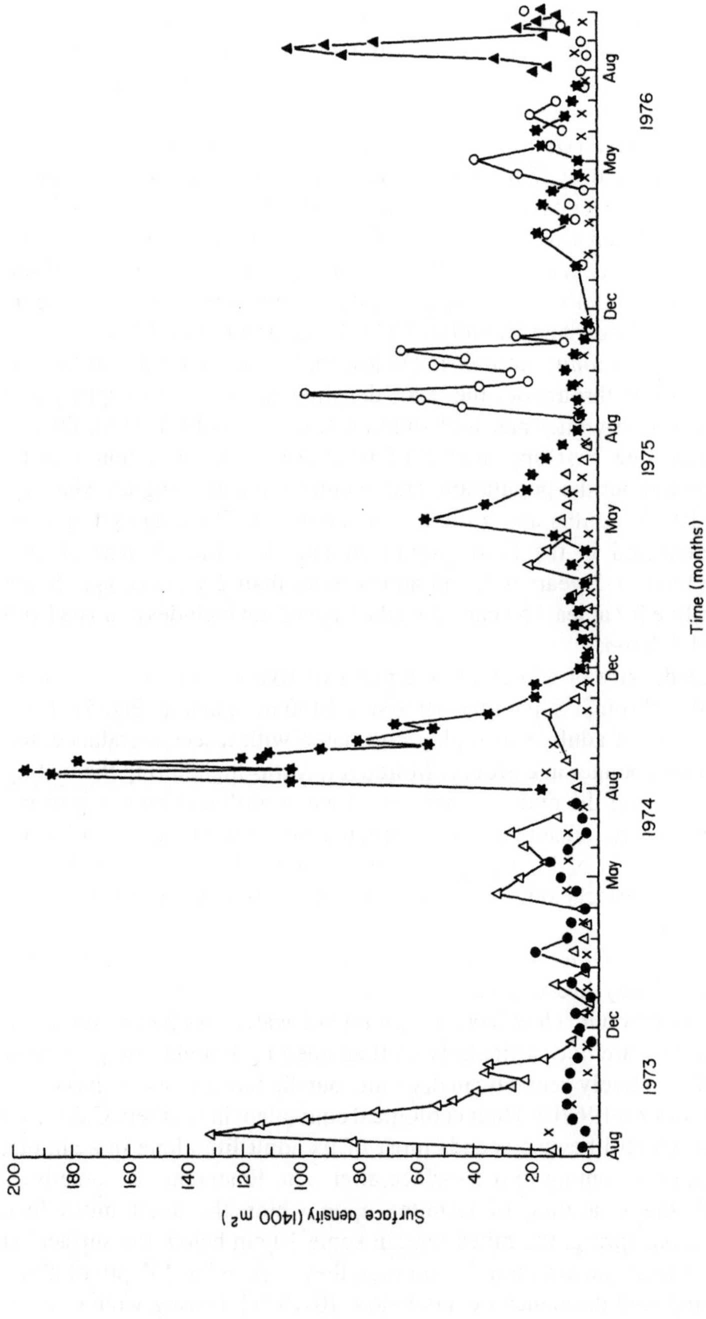

Fig. 51. Average bimonthly surface density of adults and members of five cohorts of *Paruroctonus mesaensis* in a 1400 m^2 grid. *Symbols* show animal's year of birth: ● 1972; Δ 1973; * 1974; ○ 1975; ▲ 1976; *x* animals more than 2 years old (Polis 1980a)

6.4.2 Reptiles

Few authors have studied seasonal variations in the body composition of the reptiles that inhabit desert environments (Gilles-Baillien 1974; Gregory 1982). In general, periods of drought are characterized by progressive loss of weight, mainly due to transpiration, and an increasing concentration of electrolytes in the plasma. Many species aestivate, either continuously or sporadically: in some, aestivation is obligatory, in others, faculative. The factors that stimulate its onset, however, are as yet unknown (Gregory 1982).

Al-Hussaini et al. (1985) found that *Agama stellio* in the Western desert of Egypt has a lower blood glucose concentration, higher glycogen content of the liver and gastrocnemius muscle and higher levels of serum esterified fatty acids in winter than in summer. They concluded that in winter, when food is scarce, energy is obtained from fat reserves. Adjovi (1981) evaluated glucose-induced hyperglycemia during the annual cycle of the Sahelian monitor lizard *Varanus exanthematicus*. She found that lizards captured during October, the period of nourishment, showed slower cellular glucose assimilation than did animals collected in December, the beginning of the period of food shortage, or in February the middle of the season of starvation. Lizards captured during June, the time of transition, showed variable responses.

Seasonal acclimatization to temperature is known to occur among many invertebrate taxa including isopod crustaceans (Sect. 6.4.1), but is not well established among desert reptiles, although it almost certainly takes place. For example, Mayhew (1962) found that the mean body temperatures of *Sceloporus orcutti*, which lives in association with granite outcrops in southern California, are several degrees lower in January and February that at other times of the year. Although this might be related to the cooler weather prevalent in winter, there is some evidence that preferred temperatures are also lower and that *S. orcutti* acclimatizes upon emergence from hibernation (Mayhew and Weintraub 1971). Winter aggregations occur in a variety of reptilian species, including ones that inhabit arid and semi-arid environments.

Like other desert lizards, *Callisaurus draconoides* in Death Valley, California, becomes active between February and October, with maximum numbers abroad during summer (May-July). Activity is closely correlated with the phenology of cresote bushes. It is unimodal in spring and autumn, and bimodal in summer, although activity does not cease at midday. In this way, *C. draconoides* differs from other lizards of the area and is able to utilize an essentially empty temporal niche by being active when potential predators and competitors are in hiding (Kay 1972). Seasonal reproduction, thermoregulation, predation and other aspects of the ecology of this species have been described by Pianka and Parker (1972) and of *Uta stansburiana* by Parker and Pianka (1975); seasonal variations in the thermoregulatory behaviour and body temperature of day-active lizards of the Kalahari by Huey et al. (1977).

Even though activity is limited to warm days, *Holbrookia propinqua* is active throughout the year in the arid South Padre Island, Texas. During the summer months, a bimodal pattern of daily activity is maintained: at other times activity is unimodal (Judd 1975). Likewise, winter activity in *Tropidurus peruvianus* is unimodal and correlated with low ambient temperatures (Huey 1974; Sect. 3.4.2). *Crotaphytus reticulatus*, however, which is active in summer in Mexico and Texas, spends the winter dormant in a shallow retreat (Montanucci 1971).

In many studies of reptiles that inhabit temperate climates, increased thermal resistance has been found to coincide with increased photoperiod. For instance, Kosh and Hutchinson (1972) found significant daily cycles in the CTM of *Anolis carolinensis* acclimatized to 25 °C and different photoperiods. Parietalectomized lizards had significantly lower tolerances than controls; exposure to a 15-25 °C daily cycle resulted in a significantly higher daily mean CTM than acclimatization to a constant 25 °C at the experimental photoperiods. Photoperiod, however, may be less important in tropical and subtropical deserts where day length varies little during the year.

The nature of reptilian hibernation is somewhat of an enigma because the criteria which characterize this state in ectothermic vertebrates have never been clearly defined. In the case of *Sauromalus obesus*, however, Case (1976) found a lower preferred body temperature and a greater reluctance to bask in winter than in summer. He concluded that winter inactivity in these desert lizards resembles true hibernation. Even in their retreats, however, reptiles are able to some extent to regulate body temperature by changing their positions (Case 1976; Schall 1977).

The Australian agamid *Amphibolurus muricatus* has a mean voluntary maximum temperature (V.Max.) of 34.5-38.5 °C, depending upon the season. Animals caught in autumn have a lower V.Max. than those caught in spring and summer. This is not influenced by photoperiod or by thermal acclimation in the range 20-30 °C although it is correlated with the threshold for panting (Heatwole and Firth 1982). The mechanism for the seasonal change in V.Max. is unknown. It could be controlled by an endogenous circannual rhythm; alternatively, it might be caused by changes in the external environment but, although parietalectomy lowers the panting threshold in *A. muricatus* during spring and summer (Firth and Heatwole 1976), in other species of lizards, it raises the preferred temperature (Hutchinson and Kosh 1974; Engbretson and Hutchinson 1976; Roth and Ralph 1976).

Adult western banded geckos, *Coleonyx variegatus*, are active from April until the end of October; but immature animals are active throughout the year, although only intermittently from November to March (Parker 1972). The distances travelled are occasionally extensive for such small, slow-moving reptiles. Locomotory activity is correlated with reproduction. Gravid females are found from April to September inclusive, occurring most frequently in May and June. Two or more clutches, of two eggs each, are produced annually.

6.5 Phenology of Reproduction

The study of phenology originated with Carolus Linnaeus, who kept records of the reproductive events of plants and correlated them with climatological observations. Subsequently, the subject was studied by zoologists as well as botanists, and it is now known that the reproduction of desert arthropods and reptiles is regulated by a number of factors, of which the most important are photoperiod, seasonal changes of temperature and rainfall. Many desert arthropods are univoltine and have annual generations. Among these are woodlice, a number of Orthoptera, Hemiptera, Neuroptera and Coleoptera, Solifugae, and many species of spiders. As for Diptera, Lepidoptera and Hymenoptera, according to Crawford (1981), the univoltine-multivoltine distinction is unclear in many species and needs further study. In all taxa, the situation may be complicated by the presence of diapause and variations in fecundity levels.

Life history patterns are as variable among desert arthropods as they are among the inhabitants of other biomes. Despite this, however, they are in deserts more closely related to rainfall and plant growth than they are in less extreme environments. The linkage is particularly strong in univoltine and multivoltine taxa, but is also present in long-lived herbivores, detritivores and carnivores. Long-lived detritivores tend to have long dormant periods although their food is always present, while long-lived carnivores experience lengthy periods of food shortage. Indeed, the ability to withstand starvation is clearly an adaptation of scorpions and other carnivores to the desert environment. This subject has been discussed in some detail by Crawford (1981).

As in arthropods, seasonal timing of reproduction in desert reptiles is controlled by rainfall, photoperiod and temperature. Another influence which has been implicated in the phenological responses of desert reptiles is wind. Apart from Jackson's (1974) study of two species of *Sceloporus*, however, this factor has not been studied quantitatively.

The features of reptilian reproductive cycles and their environmental control have been reviewed by Duvall et al. (1982) who discuss the effects of temperature, photoperiod, rainfall, food supply and sexual activity. Of these, high temperature is clearly the primary environmental cue in temperate regions, but whether decreasing ambient temperatures or decreasing photoperiods are stimulatory in deserts is as yet unknown. Even in cold weather, the spiny lizard *Sceloporus jarrovi* remains active while gathered in large aggregations, emerging and basking every day. In consequence of this behaviour, unique in a temperate species, ovulation may take place in the autumn and the young mature in 6 months (Ruby 1977). Winter aggregations occur in a number of desert reptiles. Communal hibernation has been recorded in *Cnemidophorus tigris* in Colorado, while Powell et al. (1978) reported winter aggregations of the skink *Leiolopisma guichenoti* near Armidale, N.S.W., Australia. Their observations do not permit an assessment as to whether the response is a social phenomenon or merely a reflection of the scarcity of suitable sites.

Temperature may play a part in regulating androgen biosynthesis in *Tiliqua rugosa* (Bourne et al. 1986), and it is the most important proximate environmental cue for reproduction in many species of lizards (Licht 1972). Beuchat (1986) found that although both male and female *Sceloporus jarrovi* thermoregulate during summer on every day that is suitable for activity, pregnant females maintain a lower body temperature (32 °C) than do males or post-parturient females (34.5 °C). This shift in the preferred body temperature of the female reflects a compromise between two conflicting options, her own preferred body temperature and a lower optimum for development of the embryos. The hypothesis was examined using a Leslie matrix model of life history, the predictions of which were in close agreement (0.4 °C or less) with temperatures observed in the field (Beuchat and Ellner 1987). Later, Beuchat (1988) showed that the length of the gestation period and the size of the young depend upon the body temperature of the pregnant female.

The reproduction of the horned lizard *Phrynosoma cornutum* (Fig. 8b) begins in April in Texas and ceases in mid-July. A single clutch, averaging 29 eggs is produced each year, and the females do not mature until their second season after birth, by which time they have reached a snout-vent length of about 70 mm (Ballinger 1974). Its ecological equivalent, the Australian thorny devil *Moloch horridus* (Fig. 8a), shows a pronounced seasonal pattern of activity. During the coldest months (June and July) and the hottest months (January and February) it seldom moves, but there is a 3-month period of activity during the autumn, and a 5-month period spanning the late winter, spring and early summer, during which mating and oviposition take place. During hot summer days the animals may retreat into shallow underground burrows while at other times they spend the night and the middle of the day under the cover of desert shrubs (Pianka and Pianka 1970).

The North American desert whiptail lizard *Cnemidophorus tigris* (Fig. 9a) inhabits hollows and flat areas with sparse vegetation. Activity begins in mid-April to May with a peak in June. The ovarian cycle starts in May, with ovulation and mating in the first half of June: egg laying takes place between the end of June into July. The testicular cycle begins when the animals emerge, reaching its peak in mid-June, and regression is complete by mid-August at which time the hatchlings start to appear (Burkholder and Walker 1973).

Of two related and narrowly sympatric Lacertidae from the Namib desert, *Aporosaura anchietae*, is capable of continuous reproductive activity for prolonged periods, as is typical of tropical lizards, whereas *Meroles cuneirostris* (Fig. 48) has a reproductive cycle typical of temperate zone species, with spermogenesis in the spring and early summer, followed by autumnal regression. Differences in food between the two species may partially explain these contrasting reproductive cycles. Whereas *A. anchietae* is omnivorous, feeding on arthropods and wind-blown seeds depending on their availability, *M. cuneirostris* is an insectivore whose prey varies seasonally in abundance; there are periods when the available insect food is insufficient to support continuous breeding (Robinson and Cunningham 1978).

Although practically no rain falls in the desert regions of northern Chile, both *Tropidurus quadrivittatus* from the coast and *T. thereisoides* from the Atacma desert at 1250 m show seasonal reproductive cycles. Reproductively active females are to be found in spring and summer, but there is no distinct spermiogenetic cycle in the males of either species (Goldberg and Rodriguez 1986). In more humid regions of tropical South America, reproduction in *Tropidurus* spp. seems to be correlated with the time of rain (Vitt and Goldberg 1983).

Even though rainfall is seasonal, the reproductive cycle of the teiid *Ameiva ameiva* in northeast Brazil is not associated with it. Females reproduce throughout the year with peaks occurring in the dry season (June-August) and during the early part of the wet season (January-February). Most probably the availability of food resources is the proximal factor dictating the intensity of reproduction in continuously breeding tropical lizards (Vitt 1982). In *Platynotus semitaeniatus* and *Tropidurus torquatus*, which inhabit the same area, the reproductive pattern is similar, but more reproduction takes place in the dry season (July-November) than during the wet season (December-June) (Vitt and Goldberg 1983). In contrast, *Cnemidophorus ocellifer* produces three clutches annually and females containing eggs in their oviducts have been found in every month except June. Again, there is no apparent association between reproduction and wet-dry seasonality as reported in certain other species of lizard from the same locality (Vitt 1983b). A similar situation is found among the colubrid snakes: in three species studied, reproduction is continual while extended reproduction is apparent in two others (Vitt 1983a).

The stimulation of reproductive activity by increased environmental moisture as a proximal cue has been demonstrated experimentally in *Anolis sagrei* by Brown and Sexton (1973). When temperature and photoperiod were similar in both moist and dry environments, females under the former conditions were reproductively more advanced than those under dry conditions, and laid many more eggs. Reproduction of the snail-eating snake *Sibon sanniola* in the northern part of the Yucatan Peninsula of Mexico is correlated with the rainy season. Females ovulate between June and September, and lay between two and three eggs per clutch. Most of these hatch during October and November (Kofron 1983).

The first conclusive evidence for the existence of an endogenous circannual reproductive rhythm in reptiles was obtained by Cuellar and Cuellar (1977) who exposed *Cnemidophorus uniparens* females to controlled long- and short-day photoperiods and then obtained up to three cycles under constant conditions in the laboratory. This confirmed an earlier suggestion by Licht (1972), who predicted the existence of such a rhythm. As mentioned above (Sect. 6.4.2), a circannual rhythm would be especially valuable in hot deserts where photoperiodic changes in the day length are not particularly well marked and seasonal rainfall is erratic.

6.6 Conclusions

Rainfall is the principal climatic factor affecting biological productivity in hot deserts. It produces temporary rain pools which are colonized by phyllopod crustaceans and other aquatic arthropods whose eggs survive long periods of drought in a state of diapause. Species of both arthropods and reptiles that are active throughout the year usually show seasonal peaks in population density. These are correlated with a number of factors, including rainfall and plant growth, as is the phenology of reproduction in both these taxa. Photoperiodic responses are not as marked as in temperate regions.

7 Adaptations for Burrowing in Sand, Avoidance of Enemies and Defence

7.1 Introduction

The three factors of the desert environment that influence most the morphology and ecophysiology of its inhabitants are heat, drought and exposure to enemies. During their daily sojourn in retreats and burrows, nocturnal animals avoid the extremes of all these parameters. Even so, they do not entirely escape them, for the desert may still be hot at night, or it can be cold. It can be very dry, or flooded (Sect. 6.3), and enemies are never absent. Nor is there security in darkness, for light from the stars and moon in the clear desert sky is so bright that adaptive colouration is important at all times (Sect. 2.3).

Reduction of wings and the evolution of flightlessness in many desert-adapted species of insects, and especially of Coleoptera, reflects both the biological and the physical factors of the environment (Scholtz and Caveney 1988). In arid regions, the selective advantage of flight-based foraging, as in the case of trogid beetles, may be offset by the disadvantage of loss of water associated with this metabolically demanding activity. Whatever the reasons for flightlessness, the condition must render insects more vulnerable to enemies than they would be if they were able to fly away. Consequently, the ability to secrete themselves within relatively secure burrows has enhanced value to desert insects that are wingless.

7.2 Burrowing

During periods of inactivity, most desert arthropods and reptiles shelter in burrows, crevices in rocks or cracks in the soil. Although several Australian desert termite species construct large mounds (Fig. 31; Sect. 3.5), elsewhere, like desert ants (Fig. 13a,b), desert termites (Fig. 52) are subterranean. The amelioration of microclimate conditions in burrows has already been outlined (Sect. 3.2), and thermoregulation is readily achieved by movement within them (Hadley 1970a). In the following, attention will be drawn to morphological adaptations for digging burrows and the methods of their use.

Although the desert woodlouse *Hemilepistus reaumuri* is an obligatory burrower (Fig. 28a,b; Sect. 3.2), non-social desert insects tend to be facultative burrowers. Most Tenebrionidae, for example, either dig into soft sand, enter the

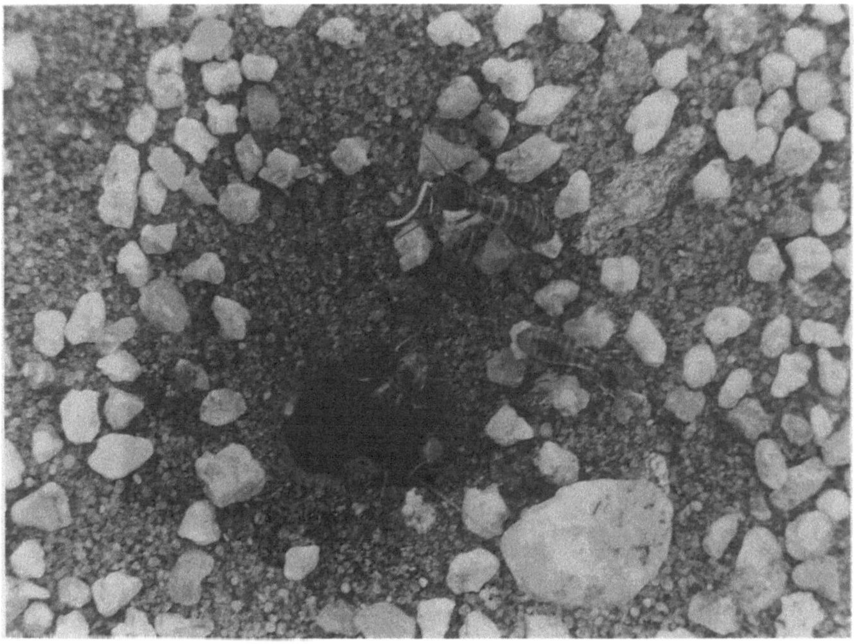

Fig. 52. Desert termites, *Psammotermes* sp., Namib desert

burrows of other animals or hide under rocks and stones when these are available. The shapes of the notum and seta are characteristic of psammophilous tenebrionid larvae, but the significance of their modifications is unknown. Adaptations specifically for digging burrows are not conspicuous among either desert insects (see Cloudsley-Thompson 1975) or reptiles.

Morphological adaptations for burrowing are, perhaps, more clearly marked among arachnids than insects. Adaptations of the scorpion *Anuroctonus phaiodactylus*, for instance, include unusually thick and heavy pedipalps, reduced metasoma and small pectines with few teeth. Digging is carried out mainly by the legs; the pedipalps being used chiefly for support. The floor of the burrow is kept smooth while the sides and ceiling are firmly packed by manipulation of the tail. Scorpions may burrow continuously for 5 or 6 h on end (Williams 1966). The burrowing habit in scorpions (Fig. 14a) is closely, but not completely, related to taxonomic position. For example, as Hadley (1974) pointed out, most members of the family Buthidae, in which the claws are slender, do not burrow. Instead, they inhabit scrapes beneath rocks and surface litter. Exceptions are provided by *Leiurus quinquestriatus* (Fig. 29) and *Parabuthus hunteri*, which normally live in holes dug deeply into the ground (Cloudsley-Thompson 1965d). In contrast, species of the families Scorpionidae, Diplocentridae (Fig. 53a,b), Vejovidae and Chactidae have massive claws and usually inhabit deep burrows which they either dig for themselves (Fig. 18a), or which have been excavated

Fig. 53. a *Nebo hierichonticus* (Diplocentridae). Oman. **b** *Urodacus manicatus* (Diplocentridae). Australia

by lizards (Fig. 18b) and small rodents. Exceptions to this general trend may reflect the role of the pedipalps in the capture of prey, courtship behaviour and defensive display (Hadley 1974).

Scorpion burrows range from simple retreats under rocks to extensive and deep holes in the ground. The Australian *Urodacus yaschenkoi* (Diplocentridae) excavates very extensive burrows which consist of an entrance, a spiral section and a terminal chamber which lies deep in the soil. Congeneric species show significant differences in the structure of their burrows (Koch 1978). Shorthouse and Marples (1980) explained how the structure of the scorpion's burrow fulfils several functions: the spiral section may minimize the chance of breaking into a neighbouring burrow; the gently sloping entrance chamber facilitates the capture of prey; the terminal chamber is situated at a depth that ensures the stability of the microclimate; while nocturnal activity results in the avoidance of many potential predators and synchronizes surface activity with that of potential prey. The burrowing activities of scorpions are discussed extensively in Polis (1990b).

Four different ecomorphotypes among scorpions are recognized by Polis (1990b): (1) lithophiles, adapted to life in cracks and crevices in rocks, have elongated, flattened bodies and appendages that allow them to use such habitats. Stout, spinelike setae operate in conjunction with highly curved claws to provide the legs with a strong grip on rough surfaces, while large, paired lateral or unguicular spines are useful in climbing; (2) psammophilic scorpions are adapted to loose sand. Long tarsal claws and enlarged setae, arranged into 'sandcombs' (Sect. 8.3.1), increase the surface area in contact with the ground. The metasoma and telson are streamlined and the pectines reduced; (3) fossorial scorpions spend almost their entire existence in burrows of their own construction; from the entrances they ambush their prey. They are often characterized by reduced telsons and large, crablike pedipalps with great crushing power; while the chelicerae, legs, setae and tarsal claws are short and robust; (4) errant scorpions move actively during foraging. They have large, slender bodies and pedipalps. Koch (1981) noted that *Urodacus* (Fig. 53b) and *Lychas* spp., in Australia, show increases in size, length of tail segments, metasomal spines, granulation, number of pectinal teeth and trichobothria, and in secondary serration of the chelicerae in correlation with increasing aridity between the coasts and central desert areas.

Despite the extensive digging operations, so necessary for their survival, Solifugae lack specific organs for burrowing (Cloudsley-Thompson 1961c, 1977b). A comparative study of the habits of North American species has demonstrated a broad similarity in behaviour. The chelicerae are used to bite the substrate, the second and, to some extent, the third pair of legs are used to rake loosened particles back under the body beneath the fourth pair of legs. The chelicerae, sometimes assisted by the palps and legs, are also used to plough excavated materials out of the burrow (Muma 1966). The burrow of *Galeodes granti* (Fig. 54) usually follows a convoluted course at a depth of 10-20 cm and often extends for more than a 1 m into the soil (Cloudsley-Thompson 1961c). Not only are burrows used as retreats during daylight hours, but they are also constructed for digestion, ecdysis and the deposition of eggs (Muma 1967). An individual solifugid of the family Eremobatidae may, during its lifetime, dig 40 or more burrows (Muma 1966).

The burrows of desert spiders not only shield their inhabitants from inclement climatic extremes, but help them to avoid the attention of predators and parasites. Such protection is only relative, however, and parasitoids such as tarantula hawks (Pepsidae; e.g. *Schistonyx* spp.; Fig. 55) are not greatly hampered by them. According to Shook (1978), the burrows of *Lycosa carolinesis* in Arizona serve as retreats from some predators. They are usually furnished with turrets reaching on average a height of 2.1 cm above the soil surface. These may serve as protection from sand or flood (Sect. 6.3). At the same time, since the sticks and

Fig. 54. *Galeodes granti* (Galeodidae), burrowing, Sudan

Fig. 55. Tarantula hawk *Schistonyx* sp., Namib desert

grasses used in their construction radiate outwards for several centimetres and are attached to the lining of the burrow, the turret may also serve as a warning system for the spider.

Carparachne aureoflava (Eusparassidae) and *Leucorchestris arenicola* (Heteropodidae; Fig. 56), the white lady spider of the Namib desert dunes, inhabit silk-lined burrows, 33 ± 13 cm in length, in soft dune sand (Henschel 1991). Without a lining of adhesive silk, however, there would be nothing to prevent the sand from falling in, and the construction of permanent burrows, with concealing trapdoor lids, would be quite impossible.

Like desert arthropods, many arid zone reptiles will occupy a variety of shelters but, in open sandy plains where no alternatives are available, many lizards dig deep burrows. Such species, however, do not appear to possess any special morphological modifications for burrowing, apart from the nasal valves which prevent sand grains from being inhaled (Sect. 5.2.1.2), hardened epidermis and scales, and strong claws (but see Sect. 8.3.3).

Fig. 56. White lady, *Leucorchestris arenicola* (Heteropodidae), Namib desert

7.3 Sand Swimming and Sand Running

In soft dune sand, permanent burrows are constructed by few arthropods other than ant-lion larvae (Neuroptera: Myrmeleontidae; Fig. 57) and some spiders (Sect. 7.2); but many dune-inhabiting arthropods and reptiles dive readily beneath the surface when they take refuge from heat or enemies.

7.3.1 Arthropods

Different behavioural, morphological and physiological adaptations in each species of fishmoths or Lepismatidae of the Namib desert dunes enable them to inhabit specific parts of the dune (Watson and Irish 1988). Thus *Ctenolepisma terebrans* and *C. pauliani* are similar in body shape to species that live commensally in human dwellings, and exploit the more stable parts of the dune environment opportunistically. In comparison with these long, thin thysanurans, *Namibmormisma muricaudata* and *Hyperlepisma australis*, which move beneath the surface of the soil, are broad at the head, with a tapered, humpbacked body, and short appendages. *Mormisma wygodzinskyi* and *Sabulepisma multiformis* have short, rounded bodies while the third pair of legs extends beyond the end of the abdomen. These species are surface-active, running rapidly over the uncon-

Fig. 57. Ant lion pits (Neuroptera:Myrmeleontidae), Qatar

solidated sand of the slip face and crest. The short, rounded bodies allow them to escape from the surface by 'swimming' into the soft slip-face sand, while the long appendages enable them to locate wind-blown detritus and to avoid predaceous spiders and Solifugae. Finally, *Swalepisma mirabilis*, whose shape is similar to that of the *Ctenolepisima* spp. but more spindlelike than dorsoventrally flattened, occupies the soft sand of the slip face. It is active during the cold, calm early mornings when advective fog moves inland. The body shape of sand dune dwellers may therefore be more related to the weather conditions under which the various species become active than it is to the medium in which they rest between their periods of surface activity (Fig. 58; Watson 1989).

In addition to Thysanura, the vegetationless dune crests and slip faces of the Namib desert also support a diverse, endemic fauna of tenebrionid beetles, arachnids and reptiles (Seely 1978). Many of these bury themselves during the heat of the day and often spend only a few hours on the surface. Breathing is possible when buried in the sand, because the quartz grains average 0.5 mm in diameter with an air-filled pore volume of 50%, and there is no dust (Sect. 5.3).

Louw et al. (1986) found that respiration in the fog-basking *Onymacris unguicularis* of the Namib desert is cyclic, with periodic bursts of oxygen uptake and carbon dioxide discharge, as has been found in many other insects (Bartholomew et al. 1985; Sect. 5.5). The carbon dioxide diffuses rapidly through the

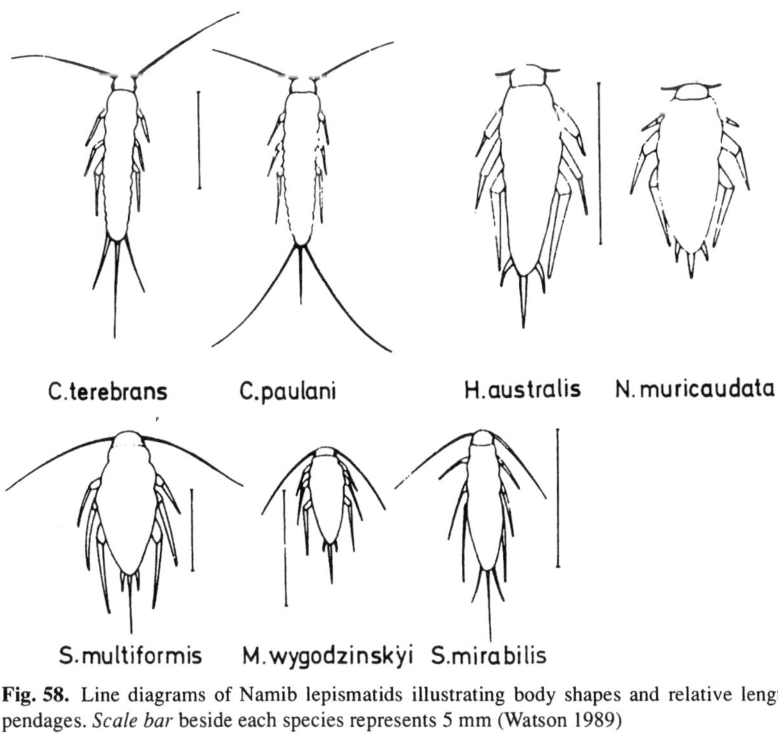

Fig. 58. Line diagrams of Namib lepismatids illustrating body shapes and relative lengths of appendages. *Scale bar* beside each species represents 5 mm (Watson 1989)

sand of the slip face, while intermittent respiration via the subelytron cavity, with long periods of apnoea, also contributes to minimizing water loss.

Ultrapsammophilous Tenebrionidae of the Namib desert include species of the genus *Lepidochora* (Fig. 59a):) of these, *L. discoidalis* (= *argentogrisea*) is the most common in the southern dune ecosystem. It is mainly confined to the leeward slip faces of the dunes where wind-blown detritus accumulates and the finest and least compact sand is found. When the animal begins to burrow, the forelegs, which are stretched forward when walking, are brought back beneath the body and excavate the sand under the head and the prothorax (Fig. 60a). These forelegs are relatively short: the broad extremities of their tibiae bear short but strong spurs (Fig. 60c). They dig the sand and pull it within reach of the middle legs. The latter, and especially the hind legs are longer, the tibial spurs are very long, digitated and spatula-shaped; the tarsal claws are also long and slightly curved (Fig. 60d). The sand is moved backwards by the middle and hind legs, whose tibial spurs and tarsi act as rakes. Tracks made during digging show the direction of sand flow (Fig. 60e,f; Coineau et al. 1982).

The discoidal shape, with the body becoming progressively thinner at its periphery (Fig. 60b), facilitates penetration into the sand, as well as change of direction within the substrate. It is suggested that the scales with which the body is covered may protect the integument from abrasion, while the longer setae protect the articulation between the prothorax and mesothorax against the intrusion of sand grains. The long bristles under the outer margin of the elytra keep sand from the ventral side of the animal while it burrows, without hampering the movements of the tarsi (Coineau et al. 1982). The general shape of the body is characteristic of beetles which swim in the sand, that is, those that dig without constructing burrows. It could also explain the shape of the central Australian pie-dish beetle *Helaesus waiteri* (Tenebrionidae; Fig. 59b), which could well be an ecological equivalent of *Lepidochora discoidalis* and *L. kahani* or, at any rate, like them be adapted for sand swimming — even though it is by no means confined to regions of dunes and loose sand. Alternatively, it may be adapted for resting in slight hollows with the edge of the body resting flat on the surface of the soil so that predators cannot get a grip on it.

The Namib desert is also home to the fastest pedestrian insect known, *Onymacris plana*, whose average running speed is 90 cm s^{-1}. In comparison, the speeds of *Physademia globosa* and *Epiphysa arenicola* are 23 and 3 cm s^{-1} respectively. These speeds are related to the ecology of the different species. *O. plana* is ultrapsammophilous, living on the slopes of the dunes in association with sparse perennial plants which provide it with food, shelter and shade. It runs swiftly over the hot sand from one plant to another (Sect. 4.2.2.1), without its body temperature becoming elevated. *P. globosa*, on the other hand, avoids extensive activity in the open, and its riverbed habitat offers shade and a variety of microenvironments. Body temperatures of this species are lower than those of *O. plana*. In contrast to both, *E. arenicola* is nocturnal and does not appear to thermoregulate behaviourally (Nicolson et al. 1984). Its hard cuticle may be an important defence against predators (Wharton and Seely 1982).

Fig. 59. a *Lepidochora kahani*, an ultrapsamophilous tenebrionid of the Namib desert dunes. **b** *Helaesus waiteri* (Tenebrionidae) of central Australia, possibly an ecological equivalent of *L. kahani*

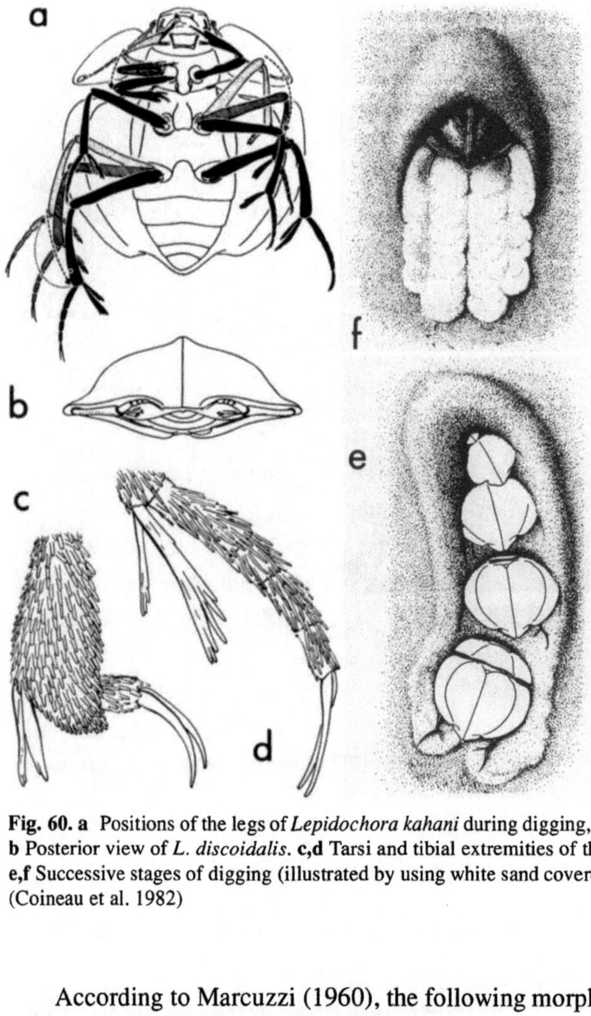

Fig. 60. a Positions of the legs of *Lepidochora kahani* during digging, drawn from film and from life. **b** Posterior view of *L. discoidalis*. **c,d** Tarsi and tibial extremities of the fore (**c**) and middle (**d**) legs. **e,f** Successive stages of digging (illustrated by using white sand covered by a thin layer of dark sand) (Coineau et al. 1982)

According to Marcuzzi (1960), the following morphological adaptations are generally found among deserticolous Tenebrionidae: (1) increase in the volume of the subelytron cavity; (2) an unusually dark-coloured integument; (3) fossorial legs in larvae and often adults; (4) large body size; (5) comparatively small spiracles; (6) frequent shortening and broadening of the body in larvae. However, C. Koch (1961) found them nearly all to be in opposition to his own observations in the Namib desert. This divergence emphasizes the uniqueness of the Namib (Seely 1978) which receives little rain but regular supplies of moisture in the form of fog. C. Koch (1961) also described a number of adaptive modifications of the legs of dune species. These modifications consist of hypertrophy of the claws, and of the bristles and spines in the regions that are in continuous contact with the sand. In addition, there are extensions of the gripping surfaces of the feet. In sand-swimming forms, such as *Lepidochora* spp., supplementary structures have been developed by means of fingerlike extensions of the normally spiniform

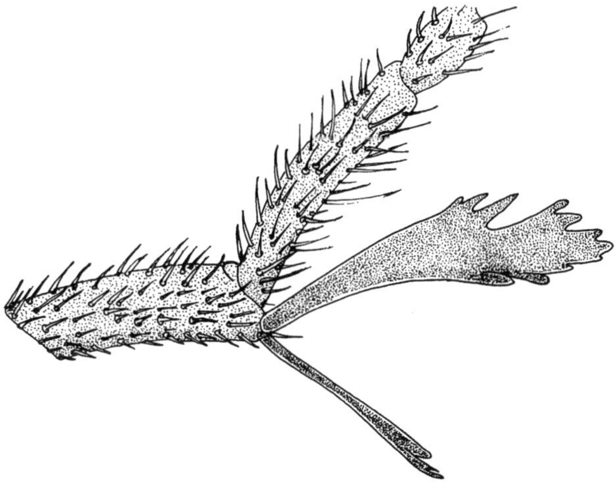

Fig. 61. Posterior leg of *Lepidochora* sp. showing apical portion of the tibia with extension of the normally spiniform calcaria and hypertrophic development of tactile bristles (Redrawn after C. Koch 1961)

calcaria (Fig. 61). An example of the diversity of body shape in the Namib tenebrionid fauna is afforded by the contrast between *Stenocara phalangium* (Fig. 62), whose long legs carry the body high above the heated surface of the sand, and *Stips stali* (Fig. 62), a strongly sclerotized nocturnal plant feeder with a contracted body, short subfossorial legs and a reduced subelytron cavity.

7.3.2 Reptiles

The lizards that inhabit sandy deserts show three principal types of morphological adaptation. They are either sand burrowers, sand swimmers (Bauer and Russell 1990) or sand runners with fringes of elongated scales on their toes (Fig. 63). The ability to enter sand is often critical for reptiles that inhabit dunes, both as a means of sheltering from the sun and of escaping predators. Sand swimmers include for example, the skinks *Scincus* spp. and *Sphenops (=Chalcides) sepsoides* in North Africa and the Middle East, iguanid lizards of the genus *Uma* and the shovel-nosed colubrid snakes *Chionactis* spp. in the Mojave desert, the lacertid *Aporosaura anchietae* (Louw and Holm 1972) and the sand viper *Bitis peringueyi* in the Namib desert. These animals all use either low amplitude, high frequency lateral undulations to sink into the sand, or else they dive headfirst into it. Because they do not construct burrows and are buried in the substrate, they usually possess morphological and physiological mechanisms for breathing beneath the sand (Stebbins 1943; Pough 1969; Huey et al. 1974).

In sand-swimming lizards and snakes, the nose or nostrum is pointed and shovel-shaped. The nostrils may be directed upwards instead of forewards, as a protection against the entry of sand. Valvelike closure of the eyes, nostrils and

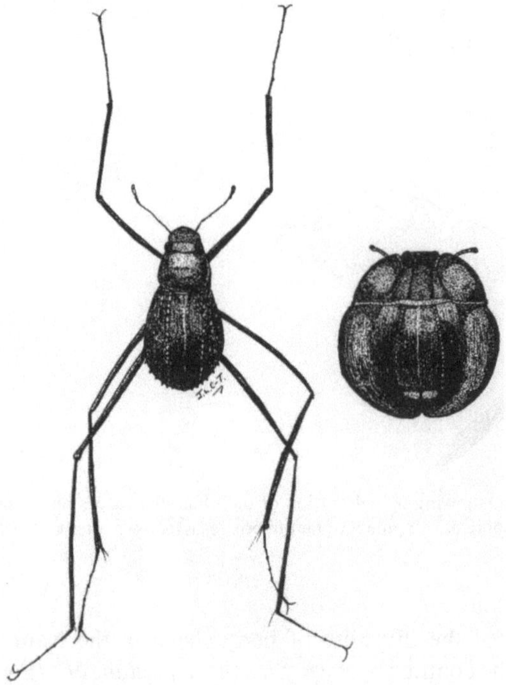

Fig. 62. *Stenocara phalangium* (*left*); *Stips stali* (*right*) (Redrawn after C. Koch 1961)

mouth enable them to dive headfirst into loose sand as though it were water. Their bodies are covered with smooth scales which cause little friction, and the legs of lizards may be reduced, or even lost, so that locomotion is accomplished entirely by wriggling. Sand burrowers, on the other hand, often have relatively short bodies and produce weak undulations, but show modifications for digging in the pes and manus. Many of them excavate discrete burrows and, consequently, do not require modifications for breathing under the sand surface. Most dune-dwelling geckos fall into this category, with the possible exception of some *Stenodactylus* spp. Sand vipers (*Echis* spp.) have scales on their backs arranged in oblique rows so that they can heap sand upon their bodies by wriggling and shaking. Horned toads (*Phrynosoma* spp.) do the same by means of lateral folds of the skin. In most burrowing snakes, the nostrils are protected by complicated valves or are reduced to small pinholes. The eyes may be overhung by head shields, or protected by scales.

Infiltration of sand into the space left by exhalation is a major problem for sand-swimming reptiles. Comparison of the respiratory movements of burrowing and non-burrowing snakes and lizards indicates that the various species have evolved different morphological and behavioural features, but the major adaptation consists of shielding the breathing movements from the surrounding sand. Thus, when *Uma notata* buries itself, it extends its forelegs backwards against

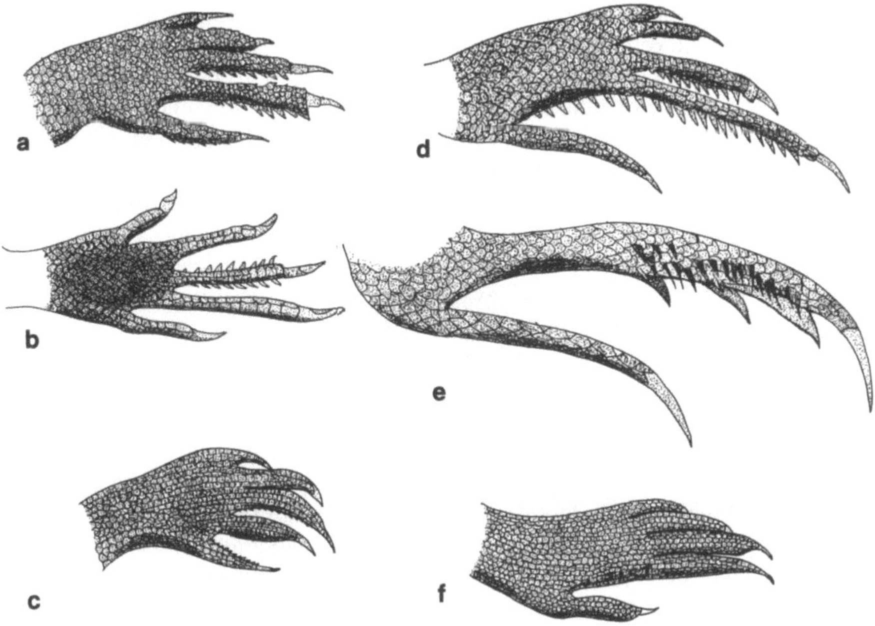

Fig. 63 a-f. Feet of desert lizards showing scales: **a-c** fore feet, **d-f** hind feet. **a,d** *Phrynocephalus* sp. **b,e** *Uma* sp. **c,f** *Stenodactylus* sp. (Redrawn after Buxton 1923)

the sides of the body so that they form a rigid roof which excludes sand, and beneath which it breathes. *Phrynosoma platyrhinos* does the same, but the rib cage is relaxed, producing folds of skin along the posterior part of the trunk which provide additional shielding (Pough 1969).

Breathing by vertical movements of the venter is a common feature of sand-dwelling reptiles. Mosauer (1932) noted that the Saharan *Sphenops sepoides* and *Scincus officinalis* have concave ventral surfaces, whereas *Chalcides ocellatus*, which inhabits the firmer sand of oases, has a rounded body. The legless lizard *Anniella pulchra* constructs open burrows in firm soil but ventures freely into dry sand (Miller 1944). It has a rounded body with a moderate ventral cavity. The burrowing sand snakes *Chilomeniscus cinctus* and *Chionactis occipitalis* also have concave ventral surfaces which Mosauer (1932, 1936) interpreted as an adaptation that prevents side-slipping in loose sand; but Pough (1969) considered that a respiratory function is more probable. The skink *Riopa sandevallii* retains the basic breathing pattern of lateral movement, but this is screened by the ridges which appear upon exhalation. The lower ridge results from the inflexibility of the ventral scales, while the stiffness of the dorsal scales and ribs contribute to the formation of the upper ridge. In consequence, respiratory movements are restricted to a narrow depression. The Indian sand boa *Eryx conicus* (Fig. 27) shows few morphological adaptations for burrowing whereas, in the larger *E. johni*, also of India, the head is pointed with a countersunk lower jaw and a pronounced cantus rostralis. *E. johni* breathes by vertical movements of the venter when it is buried, but *E. conicus* does not (Pough 1969).

The eyes of many reptiles, especially burrowing forms, are shielded by transparent 'spectacles': the lower eyelid is modified to form a window. In some species there is a single, circular transparent disc. In others, the window occupies the whole eyelid which is permanently closed and fused with the upper lid. In addition to their obvious value in excluding foreign particles from the eyes, spectacles are an important adaptation to water economy. Mautz (1980) calculated that the total evaporative water loss of five xantusiid species would increase by 5-99% if their eyes were free evaporating surfaces. Reptiles with spectacles are widely distributed in both xeric and mesic environments, however. Eye shields are clearly not an essential requirement for life in deserts: total water loss can be restricted in other ways such as closing, or partly closing the eyes whilst basking (Mautz 1982). Therefore, their primary function is probably related to sand swimming and burrowing.

Gekkonid lizards are important components of the faunas of deserts, but they are rare in dune habitats. Modifications of the feet are found in the subfamilies Diplodactylinae and Gekkoninae. Three types of adaptation are associated with dunes: fringed toes, spinose, swollen plantar surfaces and webbed feet. The first is associated with the occupation of loose, leeward dune faces, the second with the utilization of sandy substrates in general, and the third with burrowing in the compacted sand of windward dune faces. Although webs and fringes increase the surface area of the feet, and aid in locomotion on the sand, both appear primarily to facilitate movement within the substrate and in burrowing (Bauer and Russell 1991). The webbed feet of *Palmatogecko rangei* are used mainly for digging.

Lizards have evolved fringes of elongated projecting scales on their toes (Fig. 63) at least 26 times, and in 7 different families. The morphology of these fringes varies according to the type of substrate which occurs in the habitat of the species concerned. Triangular, projectional and conical fringes are associated with wind-blown sand, and these are used in similar ways: running, digging or burying in the sand. Luke (1986) discussed these modifications in terms of convergent evolution, but pointed out that not every aspect of fringe morphology has adaptive significance (see reviews by Cloudsley-Thompson and Chadwick 1964; Pianka 1985; Cloudsley-Thompson 1988b; etc.).

7.4 Adaptive Colouration

Adaptive colouration in animals has been interpreted as serving three main functions: concealment, advertisement and disguise. Desert animals are almost invariably cryptic, their colours matching that of the soil of the desert in which they live, or else they are black (Cloudsley-Thompson 1979a).

In a few cases, they many be white. Where the background colouration is white, as in White Sands, New Mexico, the function of the white colour must be crypsis. In other cases, white can be aposematic. The function of the white colour of some Namib desert beetles, such as *Onymacris bicolor*, could possibly be aposematic or it might be thermoregulatory (Sect. 4.4.2; see discussions in Cott 1940; Edmunds 1974; Cloudsley-Thompson 1979a; Seely 1985).

7.4.1 Cryptic Colouration

Many desert insects, such as grasshoppers (Fig. 64) and mantids (Fig. 65), as well as woodlice (Fig. 28a), centipedes (Fig. 14), Solifugae (Fig. 66), spiders (Fig. 67) and the majority of desert scorpions (Figs. 29, 53; McCormick and Polis 1990) are cryptic and match the colour of the substrate of the desert in which they live.

Accumulations of wax on the cuticle surfaces can serve an arthropod in many ways in addition to reducing water loss, as Hadley (1985) has pointed out in his perceptive monograph. For instance, deposits of surface wax on the Namib desert tenebrionid *Onymacris rugatipennis albotessallata* blend with the colour of the underlying sand and provide effective concealment. The camouflage of some arthropods extends beyond simple colour resemblance and obliterative counter-shading, to protective resemblance. The mantids shown in Fig. 65 not only match the trees on which they are resting; they actually look like broken twigs. Likewise, the grasshopper in Fig. 64 could easily be mistaken for one of the pebbles around it (Sect. 7.4.3).

Desert lizards and snakes that do not burrow tend often to be slender and speedy, possibly to a greater extent than species from other open habitats. This,

Fig. 64. Desert grasshopper (Orthoptera:Acrididae), Sudan

Fig. 65. Desert mantids. *Heterochaeta kumari*, Nigeria. *Inset: Blepharopsis mendica*, Sudan

however, would be difficult to assess quantitatively. Like the Arthropoda of the desert, they tend either to be desert-coloured — so that they are not conspicuous to predators — or else they are black. The function of the latter colouration is usually aposematic — except on black sands and lava flows, where it may be cryptic (Cloudsley-Thompson 1979a). Melanism is found in 7 out of 26 species of the reptiles that inhabit the Pedro Armendariz lava field of Texas (Best et al. 1983), and melanism has long been known to occur among the reptiles of the lava fields of New Mexico (Lewis 1949, 1951). Most desert lizards have cryptic colouration (Fig. 68) and many, such as *Amphibolurus pictus*, show a remarkable degree of colour change (Mayhew 1963). Venomous forms, such as *Heloderma* spp. and some, but not all, poisonous snakes, possess aposematic colouration. For instance, *Bitus caudalis* (Fig. 69) of the Namib desert, the widespread *Echis*

Fig. 66. *Prosolpuga schultzeri* (Solifugae), Namibia

Fig. 67. Huntsman spider (Heteropodidae), Oman. Two legs are missing, perhaps lost in escaping from a predator

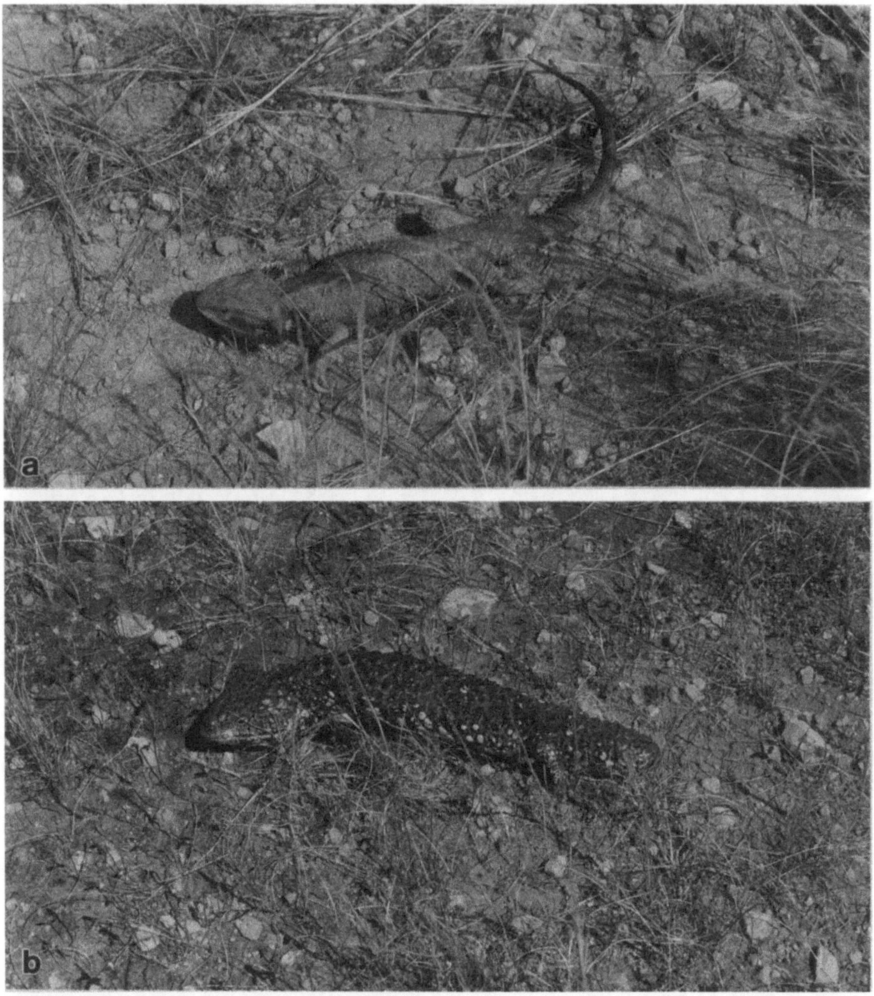

Fig. 68. **a** *Amphibolurus fordi* (Agamidae), Australia. **b** *Trachydosaurus rugosus* (Scincidae), Australia

carinatus (Fig. 70) and its subspecies of the Old World, and *Crotalus cerastes* (Fig. 11) of North America are all cryptic.

Montanucci (1978) recognized five pattern classes in *Gambelia wislizenii*, each of which predominates in populations of different physiographic regions. Those in the Great Basin desert of North America are polymorphic, but polymorphism is low in population samples from Baja California and the Sonoran and Chihuahuan deserts, while Mojave desert populations are the most polymorphic. The various pattern morphs of *G. wislizenii* appear to be cryptic in their respective habitats. White crossbands disrupt body form, an effect best suited to shrub-grass

Fig. 69. *Bitus caudalis* (Viperidae), Namibia

Fig. 70. *Echis carinatus* (Viperidae), Oman

associations; reticulate patterns are most concealing in the relatively dense *Artemsia-Atriplex* associations of the northern deserts; while punctate patterns are effective modes of concealment in relatively exposed, sandy habitats.

The rattlesnakes *Crotalus* spp. of American desert lava fields are frequently cryptically melanistic (Norris 1967; Best and James 1985). Presumably natural selection favours crypsis rather than aposematism, even though the animals are extremely poisonous, for they are cryptic elsewhere (Figs. 10, 11, 35). In addition, they appear to be unusually small in lava fields, which may be an adaptation to the size of their prey, their predators or the availability of food.

7.4.2 Aposematic Colouration

As we have already seen (Sect. 4.3.2), the preponderance of black colouration among deserticolous animals has been rationalized both in terms of maxithermy and aposematism. The problem, however, is not yet fully solved. On a yellowish, desert-coloured background, a black, aposematic wasp (Fig. 55) may be even more conspicuous than a black and yellow one. But this does not explain the presence of the black Bombyliidae (Fig. 71) to which Buxton (1923) drew attention — or to the speckled white flesh fly *Wohlfahrtia* sp. of the Namib desert, although the function here could well be one of crypsis. The universality of

Fig. 71. Desert beefly (Bombyliidae) (Redrawn after Buxton 1923)

blackness among desert beetles may well be an instance of Mullerian mimicry, since most of them have thick cuticles and appear to be distasteful to the majority of predators. Indeed, aposematic colouration, other than black, is seldom found in deserts, although Edney (1971a) suggested that the selective advantage of white elytra of *Onymacris brincki* and *Stenocara eburnea*, providing a vivid contrast to the black head, thorax and legs, may well be aposematic.

7.4.3 Mimicry

Protective resemblance, as well as Batesian and Mullerian mimicry, are common among desert arthropods (Sect. 7.4.1) and reptiles. Selection pressure is especially high in the open desert environment. A fine example of protective resemblance among cryptic reptiles is afforded by the round-tailed horned lizard *Phrynosoma modestum* (Iguanidae). From a distance this small species matches its background but, close up, it looks like a stone (Sherbrooke and Montanucci 1988). Its body size is similar to that of a small stone; its dorsal scales have a glandular texture; the centre of the back is unpatterned except for the frequent presence of small markings resembling shadow-pitting; there is visual isolation of the body from head, legs and tail by a pattern of cross-banding; the tail and legs are also dismembered by a pattern of crossbanding; the three-dimensional body form is reversed by countershading, with groin and nuchal blotches; the lizard adopts a hunchbacked attitude; and a lack of shadow-reducing fringe scales on the abdomen contrasts to their presence along the base of the tail. These adaptations strongly suggest that, unlike other members of the genus, *P. modestum* utilizes stone mimicry as a defense against visually oriented predators.

The models of Batesian mimics are not invariably distasteful, poisonous or in any way formidable. They may just not be worthwhile for a predator to attack them. For instance, in tropical regions, beetles not infrequently look like flies. In this case, the basis for the mimicry rests on the speed and maneuverability of the flies. In Central America a novel mimicry complex involves numerous taxa of beetles that resemble flies. The mimicry here appears to be primarily a Mullerian system in which agile, tree-running beetles and agile, tree-perching flies share a common colour pattern that is avoided by birds and other visually hunting predators because of the difficulty, and therefore energetic expense, in capturing them (reviewed in Cloudsley-Thompson 1980b). Similarly, in a limited area of the central Namib desert, a complex of scarab beetles, belonging to the tribe Scarabaeini, has evolved orange-coloured elytra. This aposematic advertisement is correlated with speed and agility. It has been argued that *Pachysoma denticolle* is a Batesian mimic of *Scarabaeus rubripennis* while *S. flavicornis, S. bohameni, S. rubripennis* and *Drepanopodus proximus*, but not *S. parvulus* show Mullerian speed mimicry (Holm and Kirsten 1979).

There is great flexibility between the prosoma and the opisthosoma of Solifugae. The latter can be raised until it is almost vertical, giving a remarkably scorpionlike appearance, particularly to the short-legged Rhagodidae (Fig. 72), which not only protects a vulnerable portion of the body but may well be a form

Fig. 72. *Rhagodessa melanocephala* (Rhagodidae), Sudan

of mimicry affording some protection from predators (Cloudsley-Thompson 1958). Ant mimicry is common among day-active Salticidae, especially *Myrmarachne* spp. *Cosmophasis nigrocyanea* is common in northern Sudan. The males of this species mimic ants, while the females (Fig. 73) look like mutillid wasps (Cloudsley-Thompson 1968).

Most desert lizards possess cryptic colouration — as do adult *Eremias lugubris* (Lacertidae) in the Kalahari desert — but the juveniles of this species are unusual in possessing a conspicuous pattern of black and white. In addition, they forage actively in the daytime with their backs arched, thus mimicking, in colour, gait and size, carabid beetles of the genus *Thermophilum* (= *Anthia*) (as M.K. Seely was the first to observe). The latter spray an acidic, pungent fluid when molested (Huey and Pianka 1977a). As an escape mechanism from predators, the gecko *Coleonyx variegatus* mimics a scorpion. When distrubed, the lizard usually arches its tail up over its head and, with its banded pattern, looks very much like the large scorpions (*Hadrurus* spp.) found in the same habitat. *Coleonyx* spp. do not stop moving when they are mimicking scorpions, and it is possible that their mimicry may allow some of them to escape after being cornered by predators (Parker and Pianka 1974). The incidence of broken tails is high (Parker 1972), suggesting that *Coleonyx* spp. probably suffers from a substantial amount of predation. Alternatively, this posture may merely divert predators from more vital parts of the body.

The Australian lizard *Pygopus nigriceps* has a defensive display which closely mimics the appearance of a poisonous snake. When alarmed, *P. nigriceps* elevates its head, as many snakes so, flattens its neck and, if closely approached,

Fig. 73. *Cosmophasis nigrocyanea* (Salticidae), Sudan

actively strikes at its enemy, although with its mouth closed. In a similar way, *Delma fraseri* mimicks the brown snake *Demensia textilus*, and many harmless snakes are known to mimic poisonous species (Bustard 1968c). (Mimicry has been reviewed by Cott 1940; Wickler 1968; Edmunds 1974; Cloudsley-Thompson 1980b; and many others.)

7.4.4 Warning Sounds and Threat Displays

Aposematic colouration is frequently accompanied by warning sounds; and many arthropods and reptiles, including desert species, possess sound-producing mechanisms, which usually function to deter predators. It is of no advantage to a poisonous animal to be profligate with its venom, and most will only use it as last resort. The angry buzz of mutillids, wasps and bees is by no means restricted to desert species. Stridulation has long been known to occur in scorpions of the family Scorpionidae. In *Heterometrus* spp. the sounds are made by the movement of a 'keyboard' or scraper of flattened bristles on the coxa of the pedipalp against a 'rasp' of finely striated cuticles on the coxa of the walking leg while, in *Pandinus* spp., the positions of the keyboard and rasp are reversed (Constantinou and Cloudsley-Thompson 1984). Alexander (1958) suggested that the stridulatory movements in both genera might have arisen as an elaboration of a threat display involving the pedipalps; namely that of 'intention clutching' at the threatened object. Later, she concluded that those of *Pandinus* spp. are defensive, while those described for *Heterometrus* spp. correspond to the movement of pedipalpal attack: the defensive movement causes most friction over surfaces which produce sounds in *Pandinus* spp. while attack movement does so at the points which have become stridulatory surfaces in *Heterometrus*. At least part of the anatomical bases of both mechanisms occur in another scorpionid genus, *Opisthophthalmus* (Alexander 1960).

Solifugae, like scorpions of the families Scorpionidae and Diplocentridae, are also able to make screeching sounds when irritated. In *Galeodes granti* (Galeodidae), the sound is produced by the enlarged bases of stridulatory hairs rubbing against one another; in *Rhagodessa melanocephala* (Rhagodidae) by hairs on one chelicera rubbing against ridges on the opposite chelicera and, in *Ammotrechella stimpsoni* (Ammontrechidae) by a series of ridges on one chelicera rubbing the ridges of the opposite chelicera (Cloudsley-Thompson and Constantinou 1984).

The subject of sound production among reptiles has been reviewed by Gans and Maderson (1973), who pointed out that three categories of devices are employed by reptiles. These are those involving hissing, which results from the expulsion of air from the mouth or cloaca; vocalizations or intermittent movements of air through the glottis; and mechanisms that involve rubbing or vibration of the integument. Hissing is the usual form of sound production by poisonous snakes and their mimics, although desert vipers (*Aspis* spp.) and saw-scaled vipers (*Echis* spp.; Fig. 70), in which the orientation of the lateral scales of the trunk has changed so that their keratinous keels are strongly inclined or tilted with the posterior edge downwards, stridulate, making sounds like a boiling kettle. It has been suggested that such sound production, without concomitant deep expiration, may serve to reduce respiratory water loss in desert snakes (Mendelssohn 1963).

Phrynosoma platyrhinos and *P. coronatum* have a varied repertoire of threat displays. Typical responses to predators include freezing, running and hiding. In addition, *P. platyrhinos* adopts a peculiar three-legged stance when approached by predatory leopard lizards (*Gambelia wislizenii*) and snakes. This probably causes it to appear physically larger and more difficult to swallow (Tollestrup 1981). When irritated, *P. douglassi* lowers its head so that the horns face forwards, arches its back, and hisses: *P. ditmarsi* assumes the same posture but jumps up and down, while *P. coronatum* elevates the enlarged scales on the back and stands well up on its legs. It may even lower its horns and charge the enemy. The blue tongue skink *Tiliqua scincoides* of Australia is noted for its response to enemies by swelling its body, hissing and protruding its cobalt-coloured tongue, which functions as a threat display, enhanced by rolling and folding to increase its size and effect (Carpenter 1978).

7.5 Defences

The defences of desert arthropods and reptiles are not unique but are well developed in relation to the inevitable selection pressure on animals living in extremely open country. Venomous animals and their venoms have been reviewed by numerous authors in Bucherl and Buckley (1967-71), anti-predator mechanisms of reptiles in Gans and Huey (1988).

7.5.1 Urticating Hairs

Since the time of Pliny, it has been known that the larvae of many moths and butterflies are equipped with defensive, urticating spines and hairs. Examples occur in desert regions and elsewhere (Fig. 74) and provide protection against many predators, both invertebrate and vertebrate. The effect of these hairs is not simply mechanical, but is dependent on a poisonous secretion inside them. Urticating hairs occur also in mygalomorph spiders (Fig. 75; Cooke et al. 1972), and may be a greater deterrent to would-be mammalian aggressors than the poisonous bite. Their effect, however, is wholly mechanical, and no chemicals are involved. They do not provide protection against acrocerid flies or pompiliid wasps, unlike the hairs of Dermestidae which are effective against arthropods, but their presence may explain the absence of mygalomorph spiders from the diets of lizards and snakes.

Fig. 74. Chopper caterpillar, *Gonometra postica* (Lasiocampidae), showing defensive urticating hairs, Namibia

Fig. 75. *Aphonopelma* sp. (Theraphosidae) showing defensive urticating hairs, Arizona

7.5.2 Repellent Chemicals

Tenebrionid beetles are notorious for their obnoxious smell (Kaufmann 1966). This may be due to quinones, l-alkenes and other poisonous chemicals (Tschinkel 1975). Many other arthropods also rely for defence upon the secretions of repugnant glands. These glands are well developed in large millipedes which can discharge a defensive spray for a considerable distance. The fluid contains iodine, quinine, hydrocyanic acid and small amounts of chlorine. Scolopendromorph centipedes, too, are protected not only by their venomous bites but a poison that can be dropped into the tiny incisions made by their claws. Whip scorpions (e.g. *Mastigoproctus giganteus*; Fig. 43) have glands opening at the rear of the body, while the numerous insects are also equipped with defensive glands (see Cloudsley-Thompson 1958, 1980b). A well-known example is afforded by *Eleodes* spp. (Tenebrionidae) of North America which stand on their heads when annoyed and spray their enemy with a smelly secretion that irritates the skin (Fig. 76). This substance contains benzoquinones: it is the first line of defence of these insects and has been found to repel a diversity of predators most effectively (Wickler 1968; Tschinkel 1975). No defence is perfect, however, and grasshopper mice learn to subdue quinone-secreting darkling beetles by holding them head up and forcing the tip of the abdomen downwards so that the repugnant secretion is discharged harmlessly into the soil. In addition to the secretions of defensive glands, many arthropods possess toxic chemicals in their blood and reflex bleeding occurs when they are attacked.

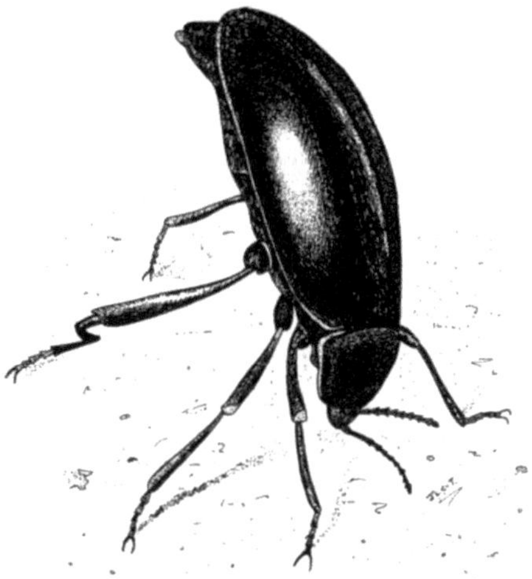

Fig. 76. *Eleodes* sp. (Tenebrionidae) spraying its defensive secretion (Redrawn after Wickler 1968)

Gopherus agassizii, and other desert tortoise species, urinate on the sand before, during and after egg laying. This softens the soil, thereby facilitating the process of digging. At the same time, it tends to camouflage the scent of the eggs. The smell of the urine, as it ages, also repels certain predators. Moreover, after the soil dries, it becomes more difficult to dig up the eggs; but the urine does not serve to inhibit bacterial or fungal growth (Patterson 1971a).

Blood-squirting from the eye has been reported on a number of occasions in the regal horned lizard *Phrynosoma solare* in Arizona. Rough handling also sometimes induces a state of trance in this and other desert lizards (for references, see Parker 1971). In addition to autotomy of the tail (Sect. 7.5.3), Australian geckos of the genera *Diplodactylus*, *Nephrurus* and *Heteronotia* possess a number of behavioural and morphological defensive adaptations which are effective in preventing or reducing predation by at least some potential predators. For instance, *D. elderi* displays towards a potential predator, the pygopod *Lialis burtonis*, by bending its body and raising its tail from which it exudes a viscid exudation. *D. williamsi* can eject a repugnant fluid for distances of 0.5 m. This readily forms cobweblike filaments which are difficult to remove and which are effective in deterring the pygmy monitor *Varanus gilleni*. *H. binoei* is protected by its heavily keeled dorsal scales, while *N. asper* lunges towards its enemies, vocalizing as it does so (Bustard 1967b, 1979).

7.5.3 Venoms

Unlike the secretions of repugnant glands, venoms that are injected by teeth or stings are usually ineffectual when merely applied to the skin or taken into the alimentary canal. They are not specific to animals which inhabit deserts and arid regions. Although many desert animals are markedly poisonous, the difference between them and their relatives of more humid regions is merely one of degree. The subject therefore will not be discussed further in the present book.

7.5.4 Autotomy

The ability to shed appendages as a method of escape from predatory enemies is known as autotomy or self-amputation. It is frequently practised by arthropods which are able to regenerate the lost appendage at the next ecdysis, but in no way is the feature especially developed among desert arthropods. In reptiles, the tail is the only organ that can be lost in this way: the anatomical process has been described by Bellairs (1969). Caudal autotomy occurs in the tuatara *Sphenodon punctatus*, in many lizards, some amphisbaenians and a few snakes. The subject has been accorded detailed reviews by Arnold (1984a, 1988) and by Bellairs and Bryant (1985) who pointed out that its adaptive value as an anti-predator device can hardly be doubted by anyone who has tried to catch lizards. It may well be especially significant for the survival of reptiles in deserts and other exposed environments. Certainly, it is a compromise adaptation involving a sacrifice that a reptile cannot easily afford but which is preferable to the alternative of almost certain death. Furthermore, loss of the tail results in a decline in social status. According to Fox et al. (1990), individuals of many species that lose their tails whilst escaping predation, incur later costs, including reduced social status. In *Uta stansburiana*, for example, removal of the tail from dominant subadult individuals lowers social status in dyadic encounters: artificial restoration of tail length restored this in females, but not in males — due largely to the dimorphic response of subordinate individuals.

The morphology of the tail, the frequency of tail loss and the positions at which tails are autotomized relative to their bases are correlated with the mode of foraging and escape from predators in many species of tropical lizards. Vitt (1983c) concluded that species which forage widely and use their running speed for escape have longer tails than most species that are sit-and-wait predators, habitat specialists, and which are cryptically coloured. He found no significant differences in the frequency of tail loss or the position at which tails are autotomized between sit-and-wait and widely foraging lizard species (Sect. 8.3.2).

Tail autotomy is an adaptive strategy of many lizards and is probably most effective in open habitats such as deserts. Congdon et al. (1974) found that rapid regeneration occurs in geckos but there is no facultative metabolic increase associated with regeneration. Energy normally allocated to growth and maintenance is devoted to regeneration of the tail. Vitt et al. (1977) investigated tail autotomy in various species of lizard from the Sonoran Desert, Arizona. They found that those which show a high frequency of tail breaks in natural populations, and in which tail autotomy is not important for escape from enemies, regenerate tails as large or larger than the originals. Species with alternative uses for the tail, on the other hand, allocate energy for regeneration to a lesser extent.

These alternative functions include swimming (e.g. *Amblyrhynchus* spp.), defence (e.g. *Ctenosaura* spp.), climbing (e.g. *Chamaeleo* spp.), balance (e.g. *Anolis* spp.), and so on. Fat storage tails may be actively functional (e.g. *Heloderma* spp.) when the fat deposit is important for survival through uncertain conditions, or passively functional — when fat deposition is a means of rapid regeneration (e.g. *Coleonyx* spp.) or a non-critical energy store (e.g. *Eumeces* spp.). In species with actively functional tails, autotomy is improbable since the loss of the tail would significantly decrease fitness. Species which use other predator escape mechanisms, but can autotomize if necessary (e.g. *Gerrhonotus* spp.), exemplify a compromise between the various conflicting advantages. In general, the effect of tail loss on individual fitness is the principal factor determining the rate of regeneration.

The escape behaviour of *Cnemidophorus tigris* of the American deserts has been quantified by Schall and Pianka (1980) who found a correlation between broken tails and population density. They suggested that sympatric whiptail lizard species differ significantly in their escape behaviour which, in turn, may be related to differences in the microhabitats selected. Ballinger and Tinkle (1979) have argued that, although the ability to lose the tail undoubtedly increases fitness in populations where predation is significant in directing the evolution of the life history (Tinkle and Ballinger 1972), the reduce growth due to tail breakage could reduce fitness through smaller clutch size due to smaller body size, or added time required for maturation. Turner et al. (1982b) analyzed the frequencies of broken tails among hatchlings, yearlings, and 2-year-old *Uta stansburiana* in southern Nevada over the years 1966-73. They found a significant correlation between the tail-break frequencies of hatchling lizards and the densities of their important predator, *Gambelia wislizenii*. There was no correlation, however, between predatory density and frequency of total breaks in older lizards.

Although the tail plays an important role in running (Vitt et al. 1977), there have been few analytical studies on the relationship between autotomy and speed in lizards. Punzo (1982), however, showed that running speed in *Cophosaurus toxanus* and *Uma notata* was considerably reduced after autotomy; a 32% decrease in *C. toxanus* and one of 42% in *U. notata*. Ballinger et al. (1979) reported a decrease of 36% in *Cnemidophorus sexlineatus* under similar conditions. It seems probable that tail autotomy may occur less frequently in species of open desert, which rely upon speed as a means of escaping from predators, than in lizards whose principal means of escape depends upon tail breakage.

The nocturnal Chihuahuan desert gecko *Coleonyx brevis* exhibits defensive behaviour involving tail display and body positioning in the presence of predatory snakes (*Hypsiglena torquata*). Autotomy occurs if an attack takes place. Tail autotomy is less frequent, and defensive displays are not found in *C. reticulatus*; but this species is considerably larger and may deter predators in other ways (Dial 1978b). The tail of the Brazilian gecko *Lygodactylus klugei* exhibits a series of characteristics associated with both tail retention and autotomy. Tails in this species are specialized for locomotion, possess specialized cutaneous glands, and have a high lipid and energy content. Nevertheless, tail loss is high in natural populations and is presumably associated with escape from predators. The importance of possessing a complete tail, both functionally and for future escape from predators, is reflected by rapid regeneration.

The threshold for autotomy in *Gehyra variegata* has been shown by Bustard (1967a) to be lowered at cooler temperatures and, in cold torpor around 4 °C, autotomy may even take place without the tail being touched. The extent of tail autotomized by gekkonid lizards has been shown by Bustard (1968a) to depend upon the animal's body temperature. Larger portions of tails are autotomized at lower than at higher temperatures. Autotomized tails are sometimes eaten by their former possessors when opportunity permits (J.L. Cloudsley-Thompson pers. observ.; C. Grant 1957) although this was not found to be the case in geckos by Congdon et al. (1974). Despite the fact that tail autotomy and tonic immobility are separate predator defense mechanisms, the two appear to be interconnected. Indeed, tonic immobility in geckos is suppressed by tail loss, after which it is, presumably, maladaptive (Herzog and Drummond 1984). Evolutionary aspects of tail shedding in reptiles have been reviewed by Arnold (1984a).

7.5.5 Flight

The desert environment has produced few specific modifications in the ways in which arthropods and reptiles flee from their enemies. Cartwheeling, as a mode of escape among sand dunes, has been recorded only among spiders. *Carparachne aureoflava* (Eusparassidae) of the Namib desert, cartwheels down smooth sand dunes when escaping from pompiliid wasps such as *Schistonyx aterrimus* (Fig. 55). Little energy is required to reach speeds of 0.5-1.5 m s^{-1} and, at 10-14 rotations s^{-1}, the spider's outline becomes blurred so that it no longer presents the normal stimulus to its adversary. Predation rather than the distribution of food may therefore influence habitat distribution in this species. Long, silk-lined burrows in steeply sloping, unconsolidated dune sand decrease a wasp's ability to excavate the spiders and increase a spider's ability to flee when excavated (Henschel 1990). Some jumping spiders (Salticidae) also cartwheel when disturbed, so it is possible that the use of the cartwheel by animals, which inhabit dunes that do not present obstacles, may be more common than is generally appreciated.

The defenses used by desert agamid lizards have been shown by Hertz et al. (1982) to change as a function of body temperature. At high temperatures, *Agama savignyi* and *A. pallida* of the Negev desert flee rapidly from predators. At lower body temperatures, which reduce their speed of sprinting, however, the lizards seldom run but instead hold their ground and attack aggressively. This temperature-dependent switch in defensive behaviour may have evolved because lizards that live in open desert habitats would stand little chance of outrunning their predators when they are cold (Hertz et al. 1982). I have observed similar responses in *Uromastix microlepis* in Kuwait. When very cold this species is dark in colour and shows a death-feigning response. On warming, however, the animals first defend themselves — exhaling with loud hisses and swinging their scaly tails

— a most effective weapon. Only when considerably warmer do they recover their natural cryptic colour and run away.

Some desert lizards, such as *Uta stansburiana*, show a direct relationship between body temperature and the avoidance of predators (Fox 1978). Others compensate for the sluggishness, induced by low body temperature, by increasing the distance that they will allow a predator to approach before they run away. Rand (1964) observed this in *Anolis lineantopus* which lacks the ability to inflict damage upon a predator. Yet another response, characteristic of *Agama* spp., *Uromastix* spp. and *Gambelia wislizenii*, is to remain stationary, display or even counterattack at lower temperatures when locomotion is impaired, and to flee rapidly when their bodies are warmer (Crowley and Peitruszka 1983).

The effect of temperature on anti-predator behaviour may indeed be critical for survival in reptiles (Greene 1988). Military dragons [*Ctenophorus* (= *Amphibolurus*) *isolepis*], found throughout the desert regions of central and western Australia, show a greater tendency to flee into shade as their body temperatures increase, even though shade is least available at midday. They tend to remain in the sun when their body temperatures are lower (Losos 1988). At high body temperatures, *Gambelia wislizenii* (Iguanidae) also flees rapidly but, at low temperatures, it adopts an aggressive posture attacking intruders noisily. Thermal effects on the escape behaviour of desert lizards are widespread. They are particularly important in open country.

Few comparable studies exist for snakes (Greene 1988) but, in some species, flight distances have been shown to vary with body temperature. Others, such as *Crotalus viridis*, show a greater propensity to strike when their bodies are cool, and to flee when warm (Goode and Duvall 1989).

Female prairie rattlesnakes, *Crotalus viridis*, usually remain immobile, relying on camouflage to avoid detection from enemies. After giving birth, however, they rattle much more readily and even coil up in the strike position when approached, rather than slipping down the nearest crevice in the soil as quickly as possible, which is their normal response (Graves 1979).

7.5.6 Sensory Physiology

Undoubtedly a response to heavy selective pressure from predation in an exposed environment, the senses of desert reptiles tend to be especially acute. A comparable situation is not to be expected in desert arthropods because they are so much smaller that they would not be affected by the macro-environment. Desert tortoises (*Gopherus* spp.) are able to perceive deep cliffs visually and by tactic sensing with their forelimbs. In this way they avoid falling into crevices in the open desert (Patterson 1971b). *Crotaphytus reticulatus*, which inhabits thorn bush deserts of Texas and Mexico, is particularly wary, seeking refuge in brush piles, *Opuntia* thickets, pack-rat nests and rodent burrows when disturbed (Montanucci 1971).

The homing ability of *Dipsosaurus dorsalis* is exceptionally well developed, lizards having returned to their homes after removal to distances of 250 m. These distances exceed those previously reported in lizards — even in other desert-dwelling species (Krekorian 1977). Similarly, *Varanus* spp. of the Great Victorian desert are exceedingly wary and essentially unapproachable (Pianka 1982). Of significance in this context is the discovery of Werner (1972b) that the effects of temperature on the function of the inner ear of various iguanid lizard species are correlated with their temperature preferences. Thus, optimal sensitivity for cochlear potentials corresponds well with the specific preferred body temperature as determined by physiological ecologists, and is higher in heliotherms than in non-heliotherms. Of course, none of the protective devices mentioned above are effective against all enemies. An unusual example is afforded by the fact that both *Dipsosaurus dorsalis* and *Coleonyx variegatus* are known to have been attacked and eaten by tenebrionid beetles when trapped (Banta 1962).

Snakes are sensitive to vibrations of the substrate and, in some cases, to sound. They possess two sensory systems that respond both to air-borne sound and to substrate vibrations, as shown by extracellular, slow evoked potentials recorded in the midbrain which, however, are not remarkably sensitive (Hartline 1971). Whether desert snakes possess an especially acute sense of hearing has not been established, although this could well be the case.

7.6 Conclusions

The morphological adaptations of desert arthropods and reptiles to their environment are almost entirely modifications for digging, burrowing in, and running across, sand. All the normal responses to natural selection through predation — adaptive colouration, mimicry, and so on, are probably well developed in relation to the lack of cover. Defenses consist of urticating hairs in arthropods, repellent chemicals, venoms, autotomy and flight — the latter being considerably influenced in reptiles by body temperature. Sensory physiology is also probably unusually acute in desert reptiles, but comparative studies between desert and non-desert congeneric species — so essential for an evaluation of such conjectures — are almost entirely lacking.

8 Interspecific Relationships, Feeding Specializations and Species Diversity

8.1 Introduction

The principal ways in which animals interact become apparent when one preys upon the other, or competes with it for environmental resources. There are further aspects of synecology, of course, including parasitism, pollination, symbiosis, and so on, but they are probably less widespread and ecologically important than predation. The basis of the food webs in the desert has already been discussed (Sect. 2.5): here, we shall concentrate upon nutritional factors.

Although many specific examples of predation and competition have been investigated in arid regions, surprisingly few food webs and patterns of faunal distribution have yet been studied there. Perhaps the desert best known in this context is the Namib (Edney 1966a; Seely 1978), which has indeed received considerable attention. The composition and distribution patterns of the fauna of the Namib desert ecosystem at Gobabeb has been analyzed by Holm and Scholtz (1980) who provided notes on the biology and ecology of 137 species. They accorded special attention to the apterous arthropods which dominate the system. Among vertebrates, primary feeders include certain geckos and lizards which, on occasion, eat vegetable matter. Larger predators include *Psammophis leightoni*, which has been observed to prey on *Meroles* sp., *P. notostictus*, and *Bitis peringueyi*, which sometimes lures *Aporosaura anchietae* with the thin black tip of its tail by wriggling it above the sand surface. *B. peringueyi* normally roams over the dunes at night, in characteristic sidewinding fashion, and probably hunts *Palmatogecko rangei*. Smaller predators include the barking geckos *Ptenopus garruleus* and *P. kochi*. These feed on insects, and occasionally on vegetable matter: they are nocturnal, but also emerge on overcast days. *P. rangei*, the most abundant nocturnal reptile of the dunes, is heavily preyed upon by various vertebrates, and is also eaten by white lady spiders (*Leucorchestris arenicola*; Fig. 56). Other small reptilian predators include *Meroles cuneirostris* (Fig. 48) , *A. anchietae* and *Chamaeleo namaquenisis* (Fig. 34), which can manage even the largest and hardest dune tenebrionids, as well as *Typhlosaurus braini*, the most elongated legless lizard known.

Invertebrate consumers are much more numerous in the Namib desert. Primary feeders only emerge after rain during the growth stage of the host plants. They include floricolous flies, the nara cricket (*Acanthoproctus* sp.), Hemiptera: Heteroptera, and species of Buprestidae and Meloidae. Grasshoppers and locusts appear after rain, and feed on the annual grasses: the cryptic *Crypsicerus cubicus* is the dominant species. Termites, weevils, Tenebrionidae and several species of Lepidoptera also appear. Thysanura and many tenebrionids are detritivores and/or omnivores, and there is a diversity of arachnid predators, e.g. scorpions, Solifugae and spiders, the largest of which even prey on small geckoes. Finally, whilst most detritivores are attracted to carcasses and feed on them, coprophages are few. There are two day-active, wingless scarabs, *Pachysoma rodriguesi* and *P. denticolle*, in addition to *Scarabaeus rubripennis*. The latter is a scavenger, feeding only on dead lizards, snakes and gerbils, and breeding in the carcasses of larger animals (Holm and Scholtz 1980).

8.2 Food Selection and Consumption

There are many scattered references in the literature to the food and feeding habits of desert animals, but no reviews on the topic have so far appeared, apart from that of Reichman et al. (1979) who summarized present knowledge on the diets of desert organisms. These authors concluded that the use of the majority of items ingested by desert consumers probably depends upon the water requirements imposed by the generally dry and high temperature regimes of deserts. Thus, herbaceous material and arthropods are superior to seeds and dry vegetation because they provide more moisture without requiring the animals which feed on them to have any special adaptations for their use. As is often the case in other biomes, desert consumers frequently eat those dietary items that are most abundant. If insect prey is seasonal, for example, the lizards that feed on insects may, in their food intake, mirror the availability of the different insect species.

The food supply of phytophagous animals is provided in the dune ecosystem of the Namib desert by dead plant material. This consists mainly of fragments of grass that have been blown from elsewhere. This plant detritus collects on the crests of the dunes where it forms clumps which slide down the slip faces and accumulate at their bases. According to Kühnelt (1975), detritus is eaten mainly by psammophilic tenebrionid beetles and, to a small extent, by termites (*Psammotermes* spp.; Fig. 52). These, in turn, are preyed upon by Solifugae, and spiders such as the white lady (*L. arenicola*) which also feeds upon flies blown from the Kuiseb valley and elsewhere (Seely 1978).

8.2.1 Interspecific Competition

Competition occurs when two species, requiring the same resource, reduce this to the point at which the population of one or both of them falls below that which it would be if the other were not present. The empirical evidence for competition among desert animals has been reviewed by Wagner and Graetz (1981). Only a small fraction of desert species has been investigated and, of this, a much smaller fraction has been studied in such a way as to assess the critical parameters. Evidence of competition has been found between such disparate groups as lizards and birds, rodents and ants. This suggests that interspecific competition may well be widespread among desert animals. The data available also indicate that biotic influences may collectively be important constraints on the populations of desert species. The diversity and percentage of predators are high. There is, therefore, good reason to question the widespread assumptions that: (1) desert animals are limited primarily by the physical environment, and relatively little by biotic constraints; and that (2) desert organisms may, on the contrary, be limited primarily by predation and behavioural self-limitation, thus minimizing the exploitation of, and competition for, scarce and variable resources.

Nevertheless, according to Grodon (1988), competition for food resources does occur between the sympatric harvester ants *Novomessor cockerelli* and *Pogonomyrmex barbatus* in the Lower Sonoran desert. Early in the morning, before *P. barbatus* becomes active, *N. cockerelli* fills its nest entrances with sand. This delays the beginning of the activity period of *P. barbatus* for 1-3 h. Since *P. barbatus* colonies do not compensate for this delay by increased foraging, nest-plugging by *N. cockerelli* must inevitably decrease their foraging capacity. *P. barbatus* exhibits four different tasks when outside the nest, namely foraging, patrolling, nest maintenance and upkeep of the colony refuse pile. In undisturbed older colonies, each activity is performed by a distinct group of workers while, in younger colonies, maintenance workers also forage (Gordon 1989). Many other studies have demonstrated task fidelity in ants of various species. The lability among the American harvester ants described above, therefore, may be a response to desert conditions.

The foraging ecology of the Namib harvester ants, *Messor denticornis* and *Tetramorium rufescens*, appears to differ considerably from that of North American species. There are large seasonal fluctuations in the numbers of foragers, a peak of abundance occurring 5 months after a peak in the availability of food. A second peak coincides with a second, smaller, pulse of food later in the year. Diet ranges widely, reflecting opportunistic responses to the availability of food and, despite large interspecific differences in body size, most of the food used is actually the same for both species. This suggests that interspecific competition for food is not an important factor in the ecology of these species (Marsh 1987b). In contrast to these harvester ants, the Namib formicine *Camponotus detritus* (Fig. 37) defends its nest and foraging sites against conspecifics, the territories changing in size and shape as scale-infested plants die and the ants are forced to search for new foraging grounds. Hostility is most intense between alien colonies of the same species, becoming progressively less marked as the taxonomic difference between species increases (Curtis 1985b).

Dung beetles, *Pachysoma striatum*, of Namaqualand, South Africa, exhibit foraging behaviour that is unique among the Scarabeinae and may be another adaptation to arid conditions. They forage on dry herbivore dung. The pellets are dragged forward whilst held between the hind legs: they are not incorporated into balls and rolled backwards as is usual in dung beetles. There can be little doubt that the behaviour pattern has evolved as a result of two major factors. The first is that food is plentiful, since dry pellets remain on the ground in arid regions for very long periods and second, interspecific competition for food is non-existent since dry dung is unattractive to other scarabaeine beetles. Because of the abundance of food, intraspecific competition is minimized (Scholtz 1989).

The Namib wedge-snouted sand lizard *Meroles cuneirostris* and the sympatric *Aporosaura anchietae* avoid competition, as might be expected, by having different diets. Whereas *M. cuneirostris* feeds mainly upon arthropods, *A. anchietae* has an omnivorous diet with seeds constituting the major portion of the items consumed. *M. cuneirostris* shows an active bimodal pattern of foraging during the day, but also employs sit-and-wait tactics, often robbing *Camponotus detritus* ants of food and waiting at newly formed termite mounds. Although opportunistic feeding occurs to some extent, *M. cuneirostris* generally appears to be selective in the type of food taken (Murray and Schramm 1987).

8.2.2 Intraspecific Competition

Despite the scarcity of valuable food resources, intraspecific competition in the arid regions of the world appears to be somewhat less than elsewhere. If extrinsic influences limit the growth of a population before it becomes large enough to deplete the resources of succeeding generations, there can be no selective advantage in conserving those resources. There will, on the contrary, be selection pressures to exploit the resource in such a way as to produce the largest number of progeny. Many desert insects, such as locusts, do just this. In contrast, the moth *Cactoblastis cactorum* (Phycitidae) conserves its food, the prickly pear cacti, *Opuntia inermis* and *O. stricta*, clumps egg-sticks rather than laying them at random. The proportion of eggs wasted by clumping increases rapidly with population per unit resource, and thus tends to stabilize the numbers of moths and cacti. The mechanism occurs most often in the centre of the range of both species, and less towards its southern limit (Monro 1967).

Most desert termite populations appear to be affected by intraspecific competition. According to Wood and Lee (1971), overdispersion of termite mounds suggests that competition between colonies of mound-building species is a common occurrence. They suggest that competition between colonies of *Nasutitermes exitiosus* in Australia influences the spatial distribution of the mounds. Darlington (1982) also suggested that the overdispersed distribution of mature nests of *Macrotermes michaelseni* in Kenya was related to intraspecific aggression. This species defends its foraging area, and storage pits filled with dead termites have been found at the junction of the foraging areas of two colonies. Further examples are cited by Jones and Nutting (1989).

The leaf-cutting ant *Veromessor pergandei* displays a suite of adaptations which indicate that it faces greater intraspecific competition than does *Pogonomyrmex rugosus* (Sect. 8.2.1). Territorial defence is more pronounced, and 'foraging' serves a dual function; it is both a foraging mechanism and a territorial patrolling device: as many as 75% of workers in a column may return without seeds (Rissing and Pollock 1989). This may be correlated with the fact that *V. pergandei* prefers sandy ravine bottoms. These are patchily distributed in the Mojave desert where *V. pergandei* is most numerous. *P. rugosus* is more common in the Sonoran desert: in the Mojave, its nests are small, and appear to suffer from the limited availability of water.

Aggression is inexorably tied to territoriality, and dominance to aggression. According to Fox et al. (1981), dominant *Uta stansburiana* occupy more of the superior habitats than the subordinates. In her review of social behaviour in lizards, Stamps (1977a) pointed out that *U. stansburiana* is very aggressive in semi-arid areas of Texas, but less so in the desert areas of Nevada and California where there may be considerable overlap in home ranges. She concluded that unpredictable fluctuations in food and other resources appear to rule out specific site defence, even when other factors are favourable. Thus, *Phrynosoma cornutum* feeds on ants which, in deserts, have highly clumped distributions, yet does not defend a home range; while some individuals are even nomadic. The large, carnivorous reptiles, too, appear less aggressive and territorial in deserts than in mesic regions. *Gambelia wislizenii*, for instance, is completely non-territorial and nomadic; and there is good reason why territoriality and aggressiveness should be reduced in areas where resources are too scarce to meet the energy demands of this type of social organization (Wiens 1976). The ability to withstand long periods of starvation and to respond rapidly through reproduction when food is available is an important physiological adaptation of desert arthropods and reptiles. The evolution and dynamics of intraspecific predation have been reviewed by Polis (1981).

8.3 Feeding Specializations

Many desert insects are herbivores or detritivores while most arachnids and reptiles are omnivores or carnivores; but there are exceptions. Tortoises are primarily vegetarian as are certain lizards. Medica et al. (1975) noted that the growth of desert tortoises in Nevada is greatest following winters of high precipitation which result in increased production of winter annuals. In the spring, when rainfall is low, not only is there increased mortality among desert tortoises (*Gopherus agassizii*) in California, but the production of eggs is reduced (Turner et al. 1984a). Tortoises may, even so, do better than might otherwise be expected were they not able to obtain moisture by eating cacti. According to Nagy and Medica (1977), the ability of tortoises to eat dry plants may be related to their relatively high body water content and their tolerance of osmotic stress. Cacti may, therefore, serve as a source of water which, in turn, permits the extraction of energy from otherwise useless resources.

Chuckwallas (*Sauromalus* spp.) of southwestern North America also depend upon desert plants as their only source of energy and moisture. *S. obesus* of the Mojave desert has been studied intensively (Norris and Dawson 1964; Templeton 1964a,b; Case 1976). This large lizard (average mass, 200 g) forages mainly on annual vegetation in the spring (Berry 1974) and apparently tolerates the dry summers and years of drought primarily through behavioural and physiological mechanisms (Nagy 1972, 1973). *S. hispidus*, a much larger species (average mass,

1 kg), occurs in several islands in the Gulf of California where rainfall is even less than in the Mojave desert. It emerges at temperatures above 30 °C — some 10-15 °C higher than those reported for *S. obesus* — spends less time in basking and the variance in its body temperature is greater. Environmental adaptations of *S. hispidus* include behavioural adjustments by the lizards in their daily activity, and reliance upon the fruits of perennial cacti which enable them to maintain the intake of energy and water throughout the summer. Comparatively little effort is involved in aggression, reproduction or territoriality (Smits 1985).

8.3.1 Detection and Capture of Prey

Many sand-dwelling desert animals possess a well-developed ability to detect insect prey by the use of vibratory cues. Examples include ant-lion larvae and other insects, scorpions, lizards and snakes. In the case of the scorpion *Paruroctonus mesaensis*, Brownell (1977) and Brownell and Farley (1979a) showed experimentally that both the directions and distance of a source of vibrations are determined so accurately that a single turn and forward movement are usually enough to capture insect prey the size of a cricket walking in the sand 15 cm away. At greater distances, only the direction of the source of vibration is determined. Visual and air-borne stimuli from nearby prey do not have any influence on the scorpion's behaviour. The prey is located by compressional and surface (Rayleigh) waves in the sand. Compressional waves spread away from the source of disturbance in all directions and have motion along the axis of travel; Rayleigh waves move along the surface of the soil and are vertically polarized. Velocities are relatively low: 95-120 and 40-50 m s^{-1} respectively. For frequencies between 1 and 5 kHz, the specific attenuation factor, Q, for sand is 18. Compound slit sensilla on basitarsal leg segments of sand-dwelling scorpions respond to surface waves generated by the movements of insects up to 50 cm distant, and tarsal sensory hairs respond to higher frequency, mostly compressional wave, components of the signal.

Directions of stimulation are perceived by the scorpion by sensing which of its legs is stimulated first. Delays as small as 0.2 s elicit accurate responses, but responses are most consistent to delays of 1-2 s, roughly the time that it takes for a Rayleigh wave to traverse the span of the legs. Information from compressional waves, which does not contribute to the orientation response, enables the scorpion to judge the distance of the source of stimulation. This depends either upon attenuations of the signal, or from appreciation of the delay between the stimulation of the tarsal hairs by compressional waves and the stimulation of the slit sensillum by the slower-moving Rayleigh waves (Brownell and Farley 1979b).

The trichobothria comprise another effective sensory apparatus enabling certain scorpions to localize their prey. They are sensitive to air movements (Abushama 1964; C. Hoffmann 1967; Linsenmair 1968) and the pattern of their distribution is highly adaptive. In addition, *Androctonus australis* and *Buthus occitanus* have been shown by Krapf (1986) to find motionless prey animals which had been stung and managed to escape before dying. These are detected by chemoreceptive sensilla on the tips of the chelae, the tarsi of the walking legs, and also by the pectines whose sensilla release specific responses of searching, seizing and feeding.

The vision of Solifugae seems to be a little better than that of scorpions, but the prey is detected mainly by touch. It is grabbed by the chelicerae, sometimes assisted by the pedipalpal suckers, and moved sideways to and fro through the camel spider's jaws. Moving objects, such as the tails of scorpions, tend to be seized first (Fig. 77). This gives the appearance of intelligence but, in fact, the reaction to the movement is purely instinctive. The response of a camel spider depends both on the size and activity of its prey. Males tend to bite and let go, females bite more quickly and are more tenaceous, a policy that pays dividends in encounters with dangerous adversaries such as scorpions (Cloudsley-Thompson 1961c, 1977b).

Although it has long been known that the inner ears of snakes are sensitive to vibrations of the ground (see Hartline 1971; Cloudsley-Thompson 1980b) there have been very few studies on vibration sensitivity in vertebrates. Laboratory experiments, however, have shown that the sandfish lizard (*Scincus scincus*) of the Sahara desert can respond to and orient crickets and mealworms moving over the surface of the sand at distances up to about 15 cm. The lizards obtain directional information from the vibrations engendered by their prey, localize the insects, and emerge from the sand to capture them. When walking on the surface,

Fig. 77. *Galeodes granti* devouring *Leiurus quinquestriatus*, Sudan

S. scincus often stops and plunges its head into the sand. This aids the detection of vibrations produced by insects moving through the sand and, presumably, facilitates transmission of sand-borne vibrations to the inner ear (Hetherington 1989).

8.3.2 Foraging Tactics of Predators

The search tactics of predatory animals are believed to fall into one of three categories: (1) ambush or sit-and-wait; (2) active, widely foraging search (Schoener 1971); or (3) a prey density-dependent combination of ambushing and active searching. Tropical forest and woodland scolopendromorph centipedes, scorpions and tarantula spiders (Theraphosidae) are much less active than desert forms. They are sit-and-wait predators, and often spend several days completely motionless. What movement they do show is far less rhythmic than that of desert species and tends to take place in daylight, whereas desert scolopendras, scorpions and tarantulas are strictly nocturnal (Cloudsley-Thompson 1981b). The effect of variation in prey density on the search tactics, and mechanisms of assessing it, have been investigated in the Texan *Scolopendra polymorpha* by Formanowicz and Bradley (1987). These authors found that, after 50 min exposure to high prey density, centipedes switched from active search to ambushing, while maintaining a high rate of search behaviour, and the rate of encounter with prey appeared to be the key element in this assessment of density. When the prey density was changed from low to high, centipedes switched from active search to ambush tactics, and when prey density was changed from high to low, centipedes switched from ambush to active search within 40 min. Such behaviour, Formanowicz and Bradley (1987) concluded, may decrease the unreliability of sampling information and the risk involved in making decisions about foraging in variable environments.

The role of sit-and-wait foraging, and of the stability of the biotic environment in shaping the population biology and life history traits of *Scorpio maurus* in the Negev desert, has been examined by Shachak and Brand (1983). Using a grid system, they found that (1) dispersal is common in all age groups, taking place only after heavy rainfall when soil moisture is high; (2) dispersion from clumped to random changed as a function of age; (3) dispersal is age-specific and density-dependent; (4) a high percentage of the population enters the resting stage during the dry season. Data and model analysis led to the conclusion that *S. maurus* is confronted with a relatively predictable physical environment and an unpredictable biological one. The unpredictability of prey is countered by dispersal in the rainy season and inactivity during the dry so that more prey items are available for each active individual scorpion. Dispersal and inactivity are responses to the main constraint of sit-and-wait foraging strategy, that is, the distance between the predator and its prey.

The lack of predictability of biological resources has been largely ignored by desert ecologists on the assumption that the physical environment is the most important factor in the evolution of adaptive strategies; but burrowing herbivores and seed eaters may well be confronted with the semi-basic problem of unpredictable biological resources that faces predators. This is due to the great temporal and spatial variation of productivity in deserts (Noy-Meir 1973, 1974).

Closely related lacertid lizards of the Kalahari desert differ considerably in their foraging behaviour. Some species are relatively sedentary, sit-and-wait predators, whereas others are more active, widely foraging predators. Huey et al. (1984), who studied them, suggested that sit-and-wait foraging may reflect differences in locomotory capacity, or even differences in potential levels of activity. In particular, the sit-and-wait *Eremias lineoocellata* simply does not have the stamina to be a widely foraging lizard for more than a few minutes; even the widely foraging *E. lugubris* becomes quickly exhausted when it runs at speeds above normal. The trade-off between sprint and endurance capacity suggests that widely foraging lacertids may be too slow to be effective sit-and-wait predators. Indeed, widely foraging species eat sluggish insects such as termites, or inactive prey such as scorpions in their burrows, whereas the faster sit-and-wait species are able to capture more mobile prey. Foraging also influences the types of predators that, in turn, eat such lizards (e.g. sit-and-wait snakes). Crossovers in foraging mode thus exist at different trophic levels. Daily maintenance energy expenditures of widely foraging lizards appear to be 1.3-1.5 times greater than those of sit-and-wait species in the same habitats, but gross food gains are 1.3-2.1 times greater (Huey and Pianka 1981).

8.3.3 Nature of Reptilian Prey and Its Reproductive Consequences

There is considerable literature dealing with reptilian predators and their prey, and it is often assumed that, if predators are to exploit their prey efficiently, they must allow the prey populations to maintain a level that maximizes the number of prey that can be harvested without decreasing the prey population. In the case of *Phrynosoma cornutum* and the harvester ants (*Pogonomyrmex* spp.) upon which it feeds, Whitford and Bryant (1979) have disclosed a foraging strategy which allows maximization of the availability of prey over weeks or months rather than per hour or day. Numbers of horned lizards are regulated by the availability and productivity of their prey — which is extremely high, nearly equal to the standing crop biomass.

Lizards of arid environments include a large proportion of termites in their diets (Pianka 1969a,b, 1970a,1971a; Robson and Lambert 1980), but among agamids, ants are sometimes even more numerous (Chapman and Chapman 1964; Cloudsley-Thompson 1981a) and may constitute over 50% of the stomach prey contents by volume (Pianka 1971b). Lizards and snakes tend, in general, to be catholic feeders. Termites and ants are among the more readily available of desert insects, and Schoener (1971) has argued that a preference may be shown for ants because little energy is needed to catch them. The key aspects of feeding strategies in general include optimal diet, foraging space, foraging period and the size of the foraging group.

Whiptail lizards (*Cnemidophorus* spp.) forage for prey by probing their long snouts under twigs and taking food objects from the substrate, by chasing prey detected visually, and by climbing vegetation and flushing prey from it. These combined strategies result in a great variety of prey taxa being consumed (Medica 1967; Pianka 1970b; McGovern et al. 1984). Many potential prey animals with aposematic colouration are distasteful. This implies the ability of whiptail lizards to recognize a great range of colours (Burghardt 1977). By means of field experiments on *C. inornatus*, McGovern et al. (1984) demonstrated the importance of movement, size and shape in attracting the attention of these lizards. The influence of odour cues, which may be important in the selection of prey by lizards (Burghardt 1977), was intended to be eliminated by the use of painted models. At the same time, it seems that vision is of primary importance in foraging for prey in the open environments of arid lands.

Medica (1967) compared the ecology and niche relationship of four sympatric species of *Cnemidophorus* and reviewed earlier work on the genus. He concluded that, once again, rainfall has an influence upon the availability of food and is also reflected in food preferences. During periods of drought, ants, termites and beetles made up the bulk of the food of these lizards; in wetter years insect larvae are eaten more often. Most desert lizards are insectivorous and fairly opportunistic feeders, eating any living creature that they can overpower. *Sceloporus horridus*, for instance, apparently feeds upon any insect food that is available (Medica and Arndt 1976). *Uta stansburiana* utilizes sit-and-wait foraging tactics, and is another opportunistic feeder. The types and numbers of insects primarily eaten are related to their abundance in nature (Parker and Pianka 1975). A few desert lizards, however, enjoy restricted diets. Among these, *Phrynosoma* and *Moloch* spp. specialize on ants (Sect. 2.4). Specialists on termites include *Ctenosaura, Dipsosaurus, Sauromalus* and *Uromastix* spp. Secondary carnivores, which prey primarily upon the eggs and young of vertebrates, and on the adults of smaller species, include *Crotophytus, Heloderma, Lialis* and *Varanus* spp. (for references, see Pianka 1973).

According to Vitt and Ohmart (1974), the scaly lizard *Sceloporus magister* of Arizona uses trees and rocks for foraging and basking, taking any food available. Larger individuals utilize a greater variety of prey sizes than do smaller lizards. Whereas *Callisaurus draconoides* employs a sit-and-wait foraging strategy (Vitt and Ohmart 1977a) and *Cnemidophorus tigris* forages widely, comparisons between the two sympatric species show a dietary overlap attributable to the common exploitation of seasonally abundant prey at the interface of the foraging microhabitats of the two species (Vitt and Ohmart 1977b).

Although most lizards are generalists with respect to food, and might therefore be expected to exhibit convergence rather than divergence in sympatry, subterranean skinks of the genus *Typhlosaurus* in the Kalahari desert conflict with this theoretical expectation. Where their geographical ranges coincide, morphological and behavioural character displacement has occurred in female and immature *T. lineatus*, which reduces dietary overlap with *T. gariepensis*. This can be explained by the fact that termites are an abundant food resource throughout the year. The morphological characters measured were snout-vent length and head length which are, in turn, related to the sizes of the termite castes that form the prey (Huey and Pianka 1974; Huey et al. 1974).

The metabolic expenditure and cost of foraging in *Cnemidophorus murinus* from the arid Caribbean island of Bonaire have been found by Bennet and Gleeson (1979) to be high. Metabolic expenditure represents approximately 87% during the day. The ability of this lizard to sustain aerobic activity is limited, but foraging is restricted to aerobically supportable limits. The low metabolic rate of nocturnal desert geckos is probably an adaptation to restricted foraging ranges (Putnam and Murphy 1982).

Whereas the majority of gekkonid lizards are climbing animals which rest and move with the body oppressed and close to the substrate — even while running — cursorial ground-dwelling species usually move with the legs straight and the body elevated above the substratum, even though they often rest in an oppressed posture. The explanation has been suggested that this posture, coupled with the long legs of *Ceramodactylus* and *Stenodactylus* spp., extends their range of vision when hunting, especially as they live among semi-shifting sand dunes whose surface is often rippled by the winds (Werner and Broza 1969). The evolution of elevated locomotory postures in cursorial geckos may also have increased the importance of the tail as a stabilizing organ. This may, in turn, explain why many ground-dwelling forms show reduced autotomy (Sect. 7.5.4) and a low incidence of regenerated tails (Werner 1968). *Anolis* lizards, likewise, have a conspicuous survey posture in which they sit on a vertical tree trunk or branch with the head held below the body and pointed at the ground. Many workers have argued that this position is probably used to survey the ground for food, conspecific intruders and predators. In both males and females, the survey position is taken up significantly less frequently after feeding, however, which suggests that it is not so important for the detection of intruders or predators (Stamps 1977b).

Bennett (1982) reviewed the energetics of reptilian activity in general, while Nagy et al. (1984) examined the energetic costs associated with wide foraging (as in *Eremias lugubris*) or sitting and waiting (as in *E. lineoocellata*) in the Kalahari desert where these species are broadly sympatric. They found that, although the hourly costs of foraging were much higher in *E. lugubris* than in *E. lineoocellata*, the hourly benefits were greater still. Thus *E. lugubris* made a greater profit on its investment and grew nearly twice as fast as the sit-and-wait predator. Huey et al. (1984) found that species that sprinted quickly also had limited endurance and vice versa; while feeding can reduce locomotory capacity. The amount of movement made by *Phrynosoma solare* is related to the capture of food. Adults scour larger areas and for longer periods in search of their prey, i.e. harvester ants (*Pogonomyrmex rugosus*), and their times of locomotory activity coincide with those of the ants (Baharav 1975).

Although many snakes merely seize their prey and swallow it alive, two important specializations for immobilizing prey have arisen in ophidian evolution, namely constriction and poisoning. Even among venomous snakes, constriction is common. This may be correlated with venom of low toxicity especially to prey that have developed high resistance, with heavily armoured prey, with snakes having short fangs, and with venoms that act slowly, especially on ectothermal prey. The subject has been reviewed by Shine and Schwaner (1985).

The Australian death adder *Acanthophis antarcticus* is adapted morphologically and behaviourally to ambush foraging rather than active searching. This probably yields low feeding rates but confers high probabilities of survival. These factors have favoured the evolution of delayed maturity and a low frequency of reproduction in females. The strong convergence with viperid snakes in reproduction tactics as well as in morphology suggests that many major ecological differences between elapids and viperids may be due to differences in foraging strategy (Shine 1980). The locomotory activity of snakes, and the ability to escape from predators, are often reduced after taking large meals. On the other hand, snakes that take smaller meals may need to forage more frequently, and this enhances the risk of encountering enemies (Huey and Pianka 1981). A balance has therefore to be struck between the two types of risk involved.

Savitzky (1983) has reviewed the co-adapted character complexes among snakes for burrowing and feeding. It is believed that the origin of snakes from a saurian ancestor involved a transitional period of extreme fossoriality during which the body became attenuated, the limbs reduced and the visual system altered. It is not surprising, therefore, that the more derived snakes should have repeatedly evolved specializations for digging, a character particularly common among desert species (Sect. 7.2). These include loss or fusion of cephalic scales, reduction in the number of dorsal scale rows, modifications of the pre-maxillary bone, abbreviations of the tail and other features. The width of the head is constrained: it influences both the ability to penetrate the substrate and the exposure of the cranial bones to stress (Gans 1974).

The primary type of tooth in reptiles is a simple, pointed cone firmly attached to the underlying bone. Adaptations have taken place, however, which increase efficiency in handling prey. These adaptations are particularly evident in snakes (Savitzky 1983). One remarkable modification is the evolution of hinged teeth, which fold backwards against the jaws, in six genera of snakes, representing at least three distinct phylogenetic lineages (Savitzky 1981, 1983). Parallel evolution has taken place in the saurophagous legless lizard *Lialis burtonis* which feeds predominantly on skinks (Patchell and Shine 1986).

The availability of food, itself dependent upon rainfall, has been shown to influence growth rates in *Sceloporus merriami* by Dunham (1978). In *S. undulatus* it also influences clutch size (Tinkle and Ballinger 1972): Ballinger 1974) found that the average clutch was reduced from 10.9 to 6.8 eggs during a year of drought and, whereas 90% females normally produce two clutches only 8.7% did so in the dry year. Breeding in *Uma notata* is likewise delayed after dry winters. This species depends upon insect food; but the reproduction of *Urosaurus graceosus*, which feeds on perennial plants, is not affected (Mayhew 1966b). From a survey of the literature, Vitt and Price (1982) concluded that the clutch mass of lizards tends to be relatively high in species that use sit-and-wait foraging tactics and among species in which crypsis does not involve a streamlined shape.

8.4 Species Diversity and Competition

Pre-occupation with the rigours of the desert environments should not, as Wallwork (1982) emphasized, obscure the fact that several faunal groups have exploited the desert successfully and, as a result, have achieved a high species diversity. He cited, as examples, Collembola, Psocoptera, and Tenebrionidae among taxa whose diversities in the Chihuahuan desert approach those in mesic temperate forest soils. This was confirmed by Wallwork et al. (1985). Louw and Seely (1982), however, pointed out that species diversity increases not only with the increasing diversity but also with increasing equality of importance of the species considered. An increase in species diversity may occur as a result of increased partitioning of the niches occupied. Thus, diversity may be generated along any of three major niche dimensions: spatial, temporal and trophic (Pianka 1977).

Patterns of evolutionary development and life history are as variable among desert invertebrates as they are among the invertebrates of other biomes. At the same time, reproduction, development and the use of resources are more closely related to the timing of precipitation and plant growth in deserts than they are in more mesic environments (Crawford 1981). This linkage is particularly strong in multivoltine and univoltine arthropods, as well as in long-lived herbivores and detritivores. Paradoxically, the latter tend to have long dormant periods, although they are surrounded by abundant food throughout the year. Phylogenetic reasons may, in part, account for this; but the seasonal competence — made possible by moisture — of ingestible gut symbionts to provide digestive enzymes could, according to Crawford (1981), also be involved. In contrast, long-lived carnivores, such as scorpions (Sect. 8.4.1), experience long periods of absolute or irregular food shortage, in response to which they exhibit considerable flexibility in breeding, development rates and prey selection.

Speciation takes place in isolated regions. Thus, lizards in general, and particularly geckos, have speciated in the Namib desert where approximately 10 of the 40 known local species of Gekkonidae are endemic. In contrast, snakes are less numerous there and are consequently less well known. Approximately 40 species of snakes have, however, been recorded from Namib, the majority of which are widely distributed in the eastern escarpment and inland areas (Seely and Griffin 1986).

In the next two sections, attention will be focussed upon the distributions and abundance of scorpions, a remarkably homogeneous class of arthropods and on the distribution and diversity of reptiles, a far less homogeneous class of vertebrates. It is unfortunate that closer comparisons cannot be made between the two taxa, but the approaches of arachnologists and herpetologists have not been alike. Even so, it is interesting that there should be so much more diversity among desert lizards than there is among scorpions. It may be that scorpions have occupied arid environments much longer than lizards, which colonized them comparatively much more recently, and that their morphology is more widely applicable to the various habitat niches available.

8.4.1 Distribution and Abundance of Scorpions

The importance of scorpions as predators and stabilizers within terrestrial communities is probably not appreciated to the full, but their impressive adaptations to arid environments have attracted considerable attention (Hadley 1974, 1990). Their distribution and abundance have also been studied in considerable detail and merit special attention. Scorpion bionomics have been reviewed by Williams (1987). The distribution of scorpions may sometimes be limited by the presence of suitable microclimates, while the texture of the soil has been shown to be of critical importance to several burrowing species. These include *Urodacus* spp. (Scorpionidae; Koch 1978), *Opisthophthalmus* spp. (Scorpionidae; Lamoral 1978), *Anuroctonus phaiodactylus* (Vejovidae; Williams 1966) and *Paruroctonus utahensis* (Vejovidae; Bradley 1986); while the presence of suitable rock retreats has been proposed by Smith (1966) as a primary factor limiting the numbers of *Urodacus abruptus*.

Burrowing biology and the spatial distribution of *Paruroctonus mesaensis* have been studied by Polis et al. (1986). These authors discovered that, in their first and second years, scorpions of this species exhibit a significant association with areas of high vegetation cover and large perennial plants, while adults display more random distribution. They concluded that vegetation provides younger, more vulnerable animals with a physical refuge from predation by adults of the same species. Although this species appears on the surface of the desert throughout the year, densities are highest in spring and autumn, and lowest in mid-winter and mid-summer (Polis 1980a,b). Low levels of overall surface activity are characteristic, and foraging patterns are significantly influenced by both inter- and intraspecific predation.

In the case of *P. utahensis*, a congeneric species of short grass prairie in central New Mexico, the density of scorpions is positively correlated with vegetation characteristics and inversely correlated with the hardness of the soil. Since prey biomass levels are similar on sites with different vegetation and different scorpion density, Bradley (1988a) concluded that the density of scorpions is primarily influenced by soil characteristics as they relate to burrowing, since it is only secondarily related to vegetation through the co-variation of the latter with soil texture. This species is a strictly nocturnal sit-and-wait predator (Sect. 8.3.2) and spends only a small proportion of apparently suitable nights foraging. According to Bradley (1988b), daily variation in surface activity levels can be partially explained by climatic factors during spring and autumn, but is not related to the availability of food or to interspecific interactions with other scorpions. The overall low activity of this species is probably a response to predation pressure.

Predation is indeed a major factor influencing population density in scorpions (Polis et al. 1981; McCormick and Polis 1990). The intensity of predation, by both vertebrates and invertebrates, is enhanced by the cannibalistic proclivities of many species. In the case of *Paruroctonus mesaensis*, Polis (1980b) found that several aspects of population dynamics can be explained by cannibalism, which is potentially the most effective and homeostatic self-regulatory agent and one which can contribute to the long-term stability and persistence of populations. As long ago as 1960, the suggestion was made that predation may be a partial explanation for the nocturnal habits of many species of scorpion (Cloudsley-Thompson 1961a). Since larger scorpions frequently prey on the predators of smaller individuals, food webs in the desert often become extremely complicated (Bradley 1988a).

Under conditions of reduced interspecific competition, there is selective pressure for age groups to diverge in their use of resources. This appears to be the case with *Paruroctonus mesaensis* because there is little competition from other species of scorpion; in part, as a result of frequent intraguild predation by adult *P. mesaensis* on smaller species. Animals that develop through a wide size range during their lifetime use a broader range of resources and affect other species differently than do species that have a narrower size range during their development (Polis 1984).

Stressful desert conditions may act as selective forces in the evolution of life history tactics in scorpions (Polis and Farley 1980; Crawford 1981; Louw and Seely 1982; Polis 1988; Polis and Sissom 1990). For example, the ability of female scorpions to re-absorb nutrients from their developing embryos could, as Polis (1988) pointed out, be regarded as an evolved adaptation that allows them to 'hedge their bets' under conditions where the availability of prey is unpredictable. When food is abundant, all the embryos develop but, when it becomes scarce, limited requirements can be redistributed. The frequent cannibalism observed among newborn scorpions may therefore represent another response to unpredictable food supplies (Polis 1979, 1980b, 1981, 1984, 1990a; Polis and Farley 1979a; McCormick and Polis 1990), since most embryonic growth takes place in the spring, during the period of maximum abundance of food (Polis and Farley 1979a,b; Polis 1988).

Scorpions have evolved several different strategies by which their prey are encountered but, basically, these may be classified as (1) waiting on exposed substrates; (2) active stalking; and (3) waiting in their burrows for insects to enter. The most common strategy is to leave the daytime shelter at dusk, settle in an exposed location and wait for prey (Williams 1966).

8.4.2 Distribution and Diversity of Reptiles

Habitat selection in reptiles has been reviewed by Heatwole (1977). It is a mechanism by which competition between related species is reduced, and species diversity engendered. The deserts of Australia, compared with other deserts, are particularly rich in lizard species. Pianka (1969a) suggested that the following explanations for this: (1) lizards usurp the ecological roles played by worms, insects, snakes and mammals in the North American deserts; (2) there is greater temporal partitioning because of the milder and more constant Australian desert climate; (3) Australian desert lizards are more narrowly specialized; (4) there is greater spatial heterogeneity in the Australian deserts because of an interdigitation with other types of habitat. Another reason may lie in the nature of spinifex grasses (*Triodia* spp.). These have a distinctive growth form; tightly bunched clumps of interlacing, spiky leaves serve as protection for animals small enough to enter the clumps through the interstices between the spikes. Furthermore, the microclimate within the clumps is considerably ameliorated in comparison with elsewhere. Finally, the infertile soils of Australia result in sparse and unreliable food production, a condition unfavourable to endotherms, but more suitable for ectothermic reptiles (Heatwole and Butler 1981).

Thus, Pianka (1967, 1972, 1980, 1981) like Warburg (1965b, 1978) attributed much of the diversity of reptile species to climatic variability and the heterogeneity of the habitat. This heterogeneity in time and space enables a comparatively large number of communities of lizard species to occupy desert habitats in different parts of the world. He concluded (Pianka 1980) that guild structure may actually foster diversity in desert species by means of competitive mutualisms arising from the indirect interactions between species that belong to different guilds.

Turner et al. (1984b) found by means of capture-recapture data that numbers of the fringe-toed lizard *Uma inornata* are not correlated with the physical attributes of sand alone but, rather, with the quality of sand in dunes on the lee sides of shrubs. Variables of importance are the penetrability of the sand, its coarseness on the surface, the diameter of the grains and the number of years since destabilization of the soil surface. Increased penetrability is a positive factor, surface coarseness a negative one.

Different species of desert lizards may avoid intraspecific competition with one another, even in a comparatively simple ecological system such as the creosote bush (*Larrea divaricata*) community of the Chihuahuan desert, by temporal and spatial resource partitioning. Thus, although there is considerable spatial overlap between *Cnemidophorus trigris* and *Holbrookia texana*, no less than five different patterns of activity can be distinguished. In contrast, *Uta stansburiana, Scleoporus magister, Phrynosoma modestum, P. cornutum* and *Gambelia wislizenii* are bimodal. Individual patterns of activity may vary, however: some species are active only in the morning, others in the afternoon, morning and afternoon (bimodal), or at midday (Creusere and Whitford 1981).

Members of the Agamidae frequently occupy similar or equivalent ecological niches to those of Iguanidae (Pianka 1973, 1985). Examples of ecologically equivalent species (Sect. 2.4) include *Uromastix* spp. (Agamidae) of the Great Palaearctic desert and North American *Sauromalus* and *Cachryx* spp. (Iguanidae). The southeast Asian agamid *Liolepis* sp. resembles the North American iguanid *Amphibolurus pictus*, as does *Dipsosaurus dorsalis* in the Californian desert (Mayhew 1963). The Australian agamid *Moloch horridus* is, in many ways, the equivalent of members of the iguanid genus *Phrynosoma* in North America, while the Iranian agamid *Phrynocephalus mystaceus* and the African gecko *Geckonia chazaliae* may also have many characters in common (Mayhew 1968; Mayhew and Wright 1971).

Comparative ecology, resource utilization and niche segretation among gekkonid lizards in the southern Kalahari have been studied by Pianka and Huey (1978). Resource utilization was compared with the ecologically similar nocturnal geckos of the Western Australian desert. Microhabitat utilization was found to be similar on both continents, but Australian geckos eat a greater variety of prey taxa than do Kalahari geckos. Pairs of morphologically similar species on the two continents do not, however, necessarily converge ecologically in diet and microhabitat. Ricklefs et al. (1981) used nine measurements of head, body and appendages to make a morphological analysis of the structure of communities of lizards in various desert habitats: North America, Western Australia and Southern Africa. This confirmed patterns revealed by studies of ecological relationships in the same localities. The construction of mathematical models to understand the functioning of animal populations has been employed for several taxa but, in the case of desert reptiles, appears only to have been used for the lizard *Uta stansburiana* in Nevada by Turner et al. (1976, 1982a). These authors based their model on three density-dependent parameters: clutch frequency, clutch size and adult survival over a period of 30 years. They concluded that processes relating to egg production were modelled more effectively than those influencing survival.

The ecology of horned lizards (*Phrynosoma* spp.) has been reviewed by Pianka and Parker (1975) with special reference to *P. platyrhinos*. Certain aspects of the anatomy, behaviour, diet, temporal activity pattern, thermoregulation and reproduction set the genus apart from most other lizards. The various co-adaptations observed, complement one another to make members of the genus ecologically successful, and are paralleled by those of the Australian *Moloch horridus*. (For an illustrated account of intercontinental comparisons between desert lizards, see Pianka 1985.) Density of the flat-tailed horned lizard (*Phrynosoma m'calli*) in California is positively correlated with the density of perennial plants and the densities of the harvester ants on which it mainly feeds (Turner and Medica 1982; Sect. 9.5).

Very little is known of the ecology and spacing patterns of nocturnal geckos other than *Gehyra* spp. (Stamps 1977a). Their distribution, however, may be affected by aggression and territorial behaviour. The African house gecko *Tarentola annularis* is not aggressive intraspecifically after the breeding season, but avoids resting sites that are occupied by other geckos. Individuals tend to occupy the same site for many weeks and dispersal to uncolonized sites is slow (Cloudsley-Thompson 1972b). This agrees with Stamps' (1977a) Type 3 spacing pattern as for instance does site tenure of *Hemidactylus brookii* in West Africa. Huey and Pianka (1977b) found that two semi-arboreal skinks, *Mabuya spilogaster* and *M. striata sparsa*, are sympatric only along a narrow belt in the southern Kalahari, whereas two terrestrial species, *M. occidentalis* and *M. variegata punctata*, are broadly sympatric both with one another and with the same arboreal species. The evidence suggests that the narrow zone of sympatry reflects adaptations to geographically discontinuous aspects of the physical environment, while more intense competition between the semi-arboreal species, inferred from high niche overlap, may restrict the zone of sympatry. Food is the major parameter separating the ecological niches of North American desert lizards whereas, in the Kalahari, feeding differences are slight while differences in time and place are considerable. All these dimensions are important in separating the ecological niches of Australian desert lizards. Overlap is least in the more diverse lizard communities (Pianka 1973).

Interspecific competition may be lessened by ecological variations in desert regions, as in other ecosystems. For instance, in the Chirichua Mountains and San Simon Valley of southeastern Arizona, the rattlesnake *Crotalus scutellatus* is found in barren parts of the valleys while *C. atrox* is confined to the mesquite-tarbush association near the mountains, and *C. molossus* occurs at elevations of 1500-2700 m above sea level. It has been suggested that increased aridity during the last century may have been accompanied by an extension of the range of *C. scutellatus* (Pough 1966), but why this species should be better adapted than *C. atrox* to aridity has not yet been fully explained. Nor has any difference been noted between the times of activity or the thermal preferences of the two species. *C. atrox* is rather larger than *C. scutellatus* (1700 mm, cf. 1100 mm for large males), however, and can devour adult cottontail rabbits, a feat beyond the ability of ordinary *C. scutellatus* which eats more lizards and fewer small rodents than does *C. atrox*. Where their ranges overlap, the two species are potential competitors: although food is far from being a limiting factor, lizards are relatively more plentiful in the valley and rodents near the mountains.

8.5 Interacting Factors

Many arthropods and reptiles are able to survive in the desert only because environmental extremes are ameliorated by rock cover, vegetation or plant litter (Larmuth 1979; Cloudsley-Thompson 1962a; Crawford 1981; Wallwork 1982; etc.). Furthermore, the activities of these animals are constrained, both spatially and temporally, by the very large differences in temperature and humidity that may be experienced over distances only a few centimetres apart. In a study of four species of *Eleodes*, referred to above (Sect. 2.8), Parmenter et al. (1989) concluded that the selection of microhabitats by these beetles is substantially influenced, not by the inherent physiological capabilities of metabolism or water conservation of the species concerned but, rather, by its preference for, and tolerance of, different thermal regimes.

Among Namib desert arthropods, the average species richness of detritivores and carnivores is similar, but individual numbers and estimated total biomass are much greater for detritivores (Sect. 2.5). Diurnally and nocturnally active forms are, on average, equally distributed throughout the day and night. Nevertheless, relatively low species diversity indicates the presence of dominant taxa in most habitats (Crawford and Seely 1987). In contrast, as we have just seen (Sect. 8.4.2), species diversity among Australian desert reptiles is extraordinarily high. Differences such as these can be explained only by detailed study of interacting ecological factors whose balance and relationships vary in different habitats. Investigations of this nature have seldom been carried out in desert terrain, although some examples are given in Whitford (1986) and Schmidt (1989).

The interactions between predators and prey have sometimes resulted in long-term evolutionary adaptations of both. An example is afforded by horned lizards (*Phrynosoma cornutum*), and their prey, workers of the harvester ant *Pogonomyrmex occidentalis*. The latter possess the most lethal venom of any known species of ant in the New World, and have no meaningful predators other than horned lizards which are extremely resistant to their poison. The immunity is, in part, the result of a blood factor that specifically neutralizes the venom (Schmidt et al. 1989). Furthermore, *P. cornutum* has evolved a very large stomach and alimentary tract, which is necessary for digesting its bulky and chitinous prey (Schmidt and Schmidt 1989).

8.5.1 Partitioning of Resources by Reptiles

Even the simplest and most basic ecological relationships are enormously complicated by the interaction of diverse factors, which often exert their influence in an indirect manner. For example, lizards that live in shaded environments tend to be relatively passive to ambient conditions, whereas those living in open habitats bask and thermoregulate more precisely. At the same time, the latter are more active and wary of the approach of predatory enemies. Maximum burst speeds and running capacity were measured in six species of desert lizards by Bennett (1980) and found to have much less thermal dependence than do most other physiological processes. This suggests the existence of adaptations that maintain functional behavioural capacity over a broad range of body temperatures, which is especially important in exposed environments.

Resource partitioning in desert reptiles has been reviewed by Toft (1985) who showed that a wide variety of factors may be involved. Some interrelationships can be somewhat unexpected. For instance, as already mentioned, the spines of *Moloch horridus* are not merely a defensive mechanism, but also serve to enhance the condensation of moisture (Gans et al. 1982). From a study of *Lacerta dugesii*, which showed a variety of colours related to predation, Crisp et al. (1979) concluded that colour is relatively unimportant in the heat balance of this species. Although large *Amphibolurus nuchalis* are faster than juveniles, and maximal speeds are related to body mass, individual variations, surprisingly, are not related to the relative lengths of the limbs, body and tail (Garland 1985).

The metabolic physiology of reptiles is adapted to their behaviour and mode of existence. Thus *Sauromalus hispida*, a herbivore, rarely engages in maximal activity; i.e. short bursts of anaerobically generated energy are adequate for its needs. In contrast, *Varanus gouldii*, a predator with a large home range, which often pursues its prey, has evolved mechanisms that sustain high levels of oxygen delivery to its tissues even during the most strenuous activity, and only a small lactacid debt is incurred (Bennett 1972).

In comparison with the complex structure of those of birds and mammals, the lungs of most lizards are simple and poorly vascularized. Nevertheless, in *Varanus gouldii* which is an active carnivore and chases its prey over long distances, the lung structure is much more complex than usual among reptiles.

This species possesses a high rate of oxygen consumption during activity (Bennett 1972) and does not rely extensively upon anaerobic metabolism (Bennett 1973b). In most species of reptiles, which rely principally on anaerobic metabolism for short bursts of rapid activity, high levels of lactate are generated, oxygen debt is high and recovery slow. Whereas non-carbonic blood buffers prevent a change of blood pH in *V. gouldii, Sauromalus hispidus* sustains a decrease in blood pH during activity (Bennett 1973b). It is curious to find, nevertheless, that the capacity for oxygen extraction by saurian lungs is similar to that of mammals and birds, even at identical rates of ventilation (Bennett 1973a).

Because lizards thermoregulate largely through behavioural activities, it is to be expected that aggressive displays should interact with thermoregulatory behaviour. This has been found to be the case in *Sceloporus occidentalis* by Engbretson and Livezey (1972). Male lizards showed a mean increase of 1.3 °C when in a dominant, compared with a subordinate situation, independently of physical activity or ambient fluctuations. It was suggested that metabolic heat production via a neurohormonal mechanism might be responsible for the rise of temperature which, in turn, could increase the efficiency of aerobic metabolism and lessen the need for anaerobic respiration.

Porter et al. (1973) have shown that food, water and the thermal environment limit the activity of desert reptiles, and that the controlling limit changes with the seasons. In calculating the climatic space required by the relatively large *Dipsosaurus dorsalis*, which will tolerate a maximum body temperature of 46 °C (according to Norris 1953), Porter et al. (1973) assumed a maximum body temperature of 45 °C and a minimum of 3 °C. From the climatic space diagram they constructed, it can be seen that, if a desert iguana is in still air in full sunlight, it must move into partial shade should the air temperature exceed 30 °C: and, when the air temperature at the ground surface reaches 45 °C, it must retreat into its burrow. It may, however, prefer a behavioural limit rather than the maximal lethal limit and, in fact, normally seeks cooler habitats when the air temperature exceeds 40 °C (Porter and Gates 1969).

The adaptation of a species to the desert environment can only truly be assessed by comparison with congeneric species of similar size that inhabit temperate or moist tropical habitats. In few instances has this been done. Henzell (1982), however, concluded from experimental data that the capacity of lizards of the *Egernia whitei* species group in Australia to adapt to aridity is constrained in the more arid parts of its range by the inability to reduce both the rate of evaporative water loss below a certain level and the extent of exposure to dehydrating conditions. Likewise, the genera *Amphibolurus* and *Phrynosoma* both have representatives in arid and temperate habitats. Selection for arid adaptation in the former has not resulted in marked changes in the mechanisms regulating exposure to high heat loads (Bradshaw and Main 1968) although the rates of water loss of arid zone species are lower than those of species from temperate regions (Bradshaw 1970). Arid-adapted *Phrynosoma*, however, increase exposure by retreating to their refuges only at higher body temperatures and environmental heat loads than do temperate-adapted species (Heath 1965).

The success of the spotted lizard *Eremias guttulata* in the stony environment of the Negev desert is dependent upon several small, complimentary, morphological and behavioural adaptations. These include the conservation of water and energy, as well as high fecundity, short life cycle and small size, which enables the young especially to make use of a diversity of micro-environments (Orr et al. 1979).

8.5.2 Body Size and Biomass

From collections made in different biotopes and geographical regions, Remmert (1981) concluded that biomass, average body size and the numbers of winged insects and spiders are governed primarily by the humidity of the environment. Deserts contain relatively small numbers of larger arthropods rather than larger numbers of smaller species. These large arthropods begin as eggs that develop into relatively very small larvae. Therefore there is a tendency for parental care to increase with increasing environmental aridity as discussed previously (Sect. 3.5). In drier biotopes, circadian rhythmicity is enhanced and the smallest insects are active at night (Sect. 3.3.1) (Holm and Edney 1973). In moist years, the same biotope harbours smaller insects and spiders than it does in dry years, while predation on insects also increases in proportion to the aridity of the biotope.

By means of a removal census, Busack (1975) estimated the biomass of the lacertid lizard *Acanthodactylus pardalis* in Tunisia and calculated that the maximum sustainable population was 148 males and 231 females per hectare, a biomass of 2154 ha^{-1}, males being significantly heavier than females. But annual changes in fecundity and survival may account for large changes in population size: up to 50% have been recorded in populations of *Uta stansburiana* in southern Nevada by Turner et al. (1970).

8.5.3 Interactions Between Animals and Plants

With its extreme temperatures, low moisture and high radiation, the arid southwest of North America abounds in biotic specializations that enable many plants and animals to survive in an environment where life is often close to the limits of tolerance. For example, large columnar cacti are characteristic of the Sonoran desert. Once established, they have almost no enemies and can withstand drought and high temperature; death usually results from storms, extreme winds, freezing and mechanical damage. When they have been injured or collapse, however, the rotting stumps harbour a diversity of organisms, including several species of *Drosophila* which are attracted to the specific odours produced by the decomposition of the different species of cacti. Fogleman and Heed (1989) have reviewed the chemical interactions that evoke the host plant specificity of these cactophilic *Drosophila* spp. They concluded (1) that cactus stem chemistry is a primary determinant of host plant utilization by the few endemic *Drosophila* species and (2) that stem chemistry probably also determines the volatile patterns produced by each species of rotting cactus.

Subterranean termites are the most important organisms involved in the breakdown of organic matter on the soil surface of most deserts. This matter includes the dung of livestock and other mammals, dead leaves, grasses and annual plants. Termites play a significant role in nitrogen cycling through the consumption of organic material, and their gut symbionts fix atmospheric nitrogen. By influencing the rate of infiltration of rainfall, termites also have a major influence on the water content of the soil. The interactions between termites, the availability of nitrogen and the soil-water status have marked effects on the composition and phenology of the Chihuahuan desert plant community (MacKay et al. 1989).

In the Sonoran desert, each species of termite maintains a fairly exclusive niche through a characteristic mode of attack on its preferred selection of different types of wood. The foraging activity of several species is influenced critically by temperature, moisture and other undetermined factors, while vegetative cover determines whether or not foragers investigate any particular site (Jones and Nutting 1989).

Chew and De Vita (1980) studied ten species of Sonoran desert ants with respect to their use of solid food resources, resource overlap and functional morphology. They could be separated into a seed foraging guild (six spp.), an arthropod foraging guild (3 spp.) and an omnivore. Forage size was correlated with body length: ratios of this to mandible gape averaged 1.38 (or reciprocally 0.72). The largest overlap of forage taxa within a guild was 0.75 while mean forage overlaps ranged from 0.18 to 0.59 for seed foragers, and from 0.41 to 0.55 for arthropod foragers, suggesting that body size determines the use of resources. If such resources are limiting, then competitive interactions should, in turn, affect body size.

8.6 Conclusions

The selection of food items ingested by desert consumers is influenced by both their water content and their availability. Except in certain groups, such as North American harvester ants, interspecific competition is not great. Sympatric species usually exploit different diets and both territorial defence and intraspecific competition are less intense than in mesic environments. Sand-dwelling scorpions are able to detect and localize their prey by the vibrations set up in the substrate. The scarcity and unpredictability of prey are reflected in the predominantly widely foraging search tactics of desert predators, but crossovers in foraging mode occur at different trophic levels. Capture of prey is related to the type of food ingested and to its availability. The latter, in turn, is dependent upon rainfall which, consequently, exerts an indirect influence upon reproduction. Scorpions represent a class of terrestrial predators that is extremely homogeneous: in contrast, desert reptiles show marked diversity of species. Finally, an indication is given of the extent to which both desert arthropods and reptiles interact with their environments.

9 Final Conclusions

During the course of this book it must have become apparent that neither arthropods nor reptiles show particularly marked desert adaptations. Both groups include taxa whose ecophysiology is such that they can exist and even thrive under desert conditions, mainly by avoiding thermal extremes and by the refinement of a pre-existing ability to economize in water. Nevertheless, lizards of the Australian genus *Egernia* are unable to reduce evaporative water loss below a certain level; and the persistence of temperate-adapted taxa in the arid zone is possible only in refugia offering relief from aridity (Henzell 1982). Furthermore, selection for arid adaptation in *Amphibolurus* spp. has not resulted in marked changes in the mechanisms regulating exposure; but all species of the genus prolong exposure when subjected to high heat loads (Bradshaw and Main 1968), while the rates of water loss of arid zone species are lower than those of species from temperate regions (Bradshaw 1970). Again, arid-adapted species of *Phrynosoma* increase their periods of exposure by retreating to their refuges at higher body temperatures or heat loads than do temperate-adapted species (Heath 1965). Morphological adaptations are primarily modifications for burrowing in or running over sand, whilst no unique anti-predator devices have been evolved in the desert.

Much physiological research has been restricted to causes and effects, while feedback has been somewhat neglected: but the latter has to be taken into account if ecophysiological adaptation is to be understood properly. Every adaptation must have a function or functions, both direct and indirect. At the same time, this situation is complicated by the fact that animals are not necessarily found in environments where they thrive best, but where they compete best with sympatric species.

As far as reptiles are concerned, Bradshaw (1986) put the case even more strongly. He concluded that once a particular genetic modification becomes the common property of a taxon, use of the word 'adaptation' essentially changes to the vernacular form. The organisms are now seen to be well designed to contend with the exigencies of their particular environment through the incorporation of these genetic novelties. Fitness need not imply persistent directional selection, only the stabilizing of the selection necessary to maintain the modifications which enable a population to persist in that environment. "It is thus possible to reconcile a fruitless search for specific adaptations to desert life in reptiles with their manifestly successful radiation in such habitats", he concluded, "and see this instead as a direct consequence of their relatively modest mode of existence which imposes far fewer demands on the environment than that of endotherms such as birds and mammals". Bradshaw's approach has been oriented far more physiologically than that adopted in the present book. He lays heavy emphasis on the Australian fauna with special reference to *Amphibolurus nuchalis* (Agamidae) but, nevertheless, the conclusions he reaches are applicable to deserts everywhere.

The evolution of the diversity of life has been achieved by the natural selection of chance mutations that gave individual organisms some inheritable quality affecting positively their chances of reproducing in the particular environments to which they are exposed. In addition to such genotypic adaptations, slowly developing phenotypic adaptations occur within the lifetimes of individual organisms. These are involved in acclimatization, i.e. physiological responses to natural changes in the environment. Both play, or have played, their part in the adaptations of arthropods and reptiles to arid environments; and it is important, as far as possible, to separate them when considering the ecophysiology of these animals. At the same time, it is evident that desert species do not possess metabolic properties greatly different from those of non-desert species. For instance, assimilation efficiencies, standard metabolic rates and the activity metabolism of desert reptiles are indistinguishable from those of mesic reptiles. Both arthropods and reptiles are pre-adapted to life in arid regions, and differ from non-desert species only in the extent to which they are able to withstand the exigencies of the desert environment. They are found there because they are able not only to withstand the physical conditions of the environment, but also compete best in that environment with other species.

References

Abushama FT (1964) On the behaviour and sensory physiology of the scorpion *Leiurus quinquestriatus* (H & E). Anim Behav 12:140-153

Abushama FT (1984) Epigeal insects. In: Cloudsley-Thompson JL (ed) Sahara desert. Pergamon, Oxford (Key Environments), pp 129-144

Abushama FT, Al-Houty WA (1988) The foraging activity of subterranean termites in the Kuwait Desert. J Arid Environ 14:75-82

Abushama FT, Al-Salameen MA (1989) Temperature reactions of desert tenebrionid beetles from Kuwait. J Arid Environ 16:293-304

Ackerman RA, Seagrove RC, Dmiel R, Ar A (1985) Water and heat exchange between parchment-shelled reptile eggs and their surroundings. Copeia 1985 (3):703-711

Adjovi Y (1981) Evolution of glucose induced hyperglycemia during the annual cycle of a Sahelian lizard (*Varanus exanthematicus*). Comp Biochem Physiol 69(A):529-536

Ahearn GA (1970a) Changes in haemolymph properties accompanying heat death in the desert tenebrionid beetle *Centrioptera muricata*. Comp Biochem Physiol 33:845-847

Ahearn GA (1970b) Water balance in the whipscorpion *Mastigoproctus giganteus* (Lucas) (Arachnida, Uropygi). Comp Biochem Physiol 35:339:353

Ahearn GA (1970c) The control of water loss in desert tenebrionid beetles. J Exp Biol 53:573:595

Ahearn GA, Hadley NF (1969) The effects of temperature and humidity on water loss in two desert tenebrionid beetles, *Eleodes armata* and *Cryptoglossa verrucosa*. Comp Biochem Physiol 30:739-749

Ahearn GA, Hadley NF (1977) Water transport in perfused scorpion ileum. Am J Physiol 233:R198-R207

Alexander AJ (1958) On the stridulation of scorpions. Behaviour 12:339-352

Alexander AJ (1960) A note on the evolution of stridulation within the family Scorpionidae. Proc Zool Soc Lond 133:391 399

Alexander AJ, Ewer DP (1958) Temperature adaptive behaviour in the scorpion, *Opisthophthalmus latimanus* Koch. J Exp Biol 35:349-359

Al-Hussaini AAH, El Banhawy MA, Hussein ARME (1985) Effect of environmental temperature on some metabolic aspects of the desert lizard *Agama stellio* (L.). J Arid Environ 8:73-77

Alicata P, Caruso D, Costa G, Marcellino I, Motta S, Petralia A (1979) Richerche eco-etologiche sulla fauna delle dune costiere di Portopalo (Siracusa) II Comportamento, distribuzione spaziale e ritmi di attività di *Pimelia grossa* Fabr (Coleoptera, Tenebrionidae). Animalia 6(1/3):33-48

Amos WH (1959) The life of a sand dune. Sci Am 201(1):91-99

Andersen J, Zachariassen KE, Maloiy GMO, Kamau JM (1986) Adaptations of carabid beetles to dry habitats in East Africa. J Trop Ecol 2:127-138

Anderson RA, Karasov WH (1981) Contrasts in energy intake and expenditure in sit-and-wait and widely foraging lizards. Oecologia (Berl) 49:67-72

Anderson RV, Tracy CR, Abramsky Z (1979) Habitat selection in two species of grasshoppers. The role of thermal and hydric stresses. Oecologia (Berl) 38:359-374

Andrewartha HG, Birch LC (1954) The distribution and abundance of animals. Chicago University Press, Chicago

Applin DG, Cloudsley-Thompson JL, Constantinou C (1987) Molecular and physiological mechanisms in chronobiology — their manifestations in the desert ecosystem. J Arid Environment 13:187-197

Arnold EN (1984a) Evolutionary aspects of tail shedding in lizards and their relatives. J Nat Hist 18:127-169

Arnold EN (1984b) Ecology of lowland lizards in the eastern United Arab Emirates. J Zool (Lond) 204:329-354

Arnold EN (1988) Caudal autotomy as a defense. In: Gans C, Huey RB (eds) Biology of the Reptilia 16 (Ecology B). Defense and life history. Alan R Liss, New York, pp 235-273

Atsatt SR (1939) Color changes as controlled by temperature and light in the lizards of the desert regions of southern California. Univ Calif Publ Biol Sci 1:237-276

Avery RA (1982) Field studies of body temperatures and thermoregulation. In: Gans C, Pough FH (eds) Biology of the Reptilia 12 (Physiology C). Physiological ecology. Academic Press, London, pp 93-166

Baharav D (1975) Movement of the horned lizard, *Phrynosoma solare*. Copeia 1975(4):649-657

Baker HG, Hurd PD (1968) Intrafloral ecology. Annu Rev Entomol 13:385-414

Baker LA, Weathers WW, White FN (1972) Temperature induced peripheral blood flow changes in lizards. J Comp Physiol 80:313-323

Ballinger RE (1974) Reproduction of the Texas horned lizard *Phrynosoma cornutum*. Herpetologica 30:321-327

Ballinger RE (1977) Reproductive strategies: food availability as a source of proximal variation in lizards. Ecology 58:628-635

Ballinger RE, Tinkle DW (1979) On the cost of tail regeneration to body growth in lizards. J Herpetol 13:374-375

Ballinger RE, Nietfeldt JW, Krupa JJ (1979) An experimental analysis of the role of the tail in attaining high running speed in *Cnemidophorus sexlineatus* (Reptila, Squamata, Lacertilia). Herpetologica 35:114-116

Banta BH (1962) Beetles attacking lizards. Br J Herpetol 3:39

Bartholomew GA (1982) Physiological control of body temperature. In: Gans C, Pough FH (eds) Biology of the Reptilia 12 (Physiology C). Physiological ecology. Academic Press, London, pp 167-211

Bartholomew GA, Tucker VA (1963) Control of changes in body temperature, metabolism, and circulation by the agamid lizard, *Amphibolurus barbatus*. Physiol Zoöl 36:199-218

Bartholomew GA, Tucker VA (1964) Size, body temperature, thermal conductance, oxygen consumption, and heart rate in Australian varanid lizards. Physiol Zoöl 37:341-354

Bartholomew GA, Lighton JRB, Louw GN (1985) Energetics of locomotion and patterns of respiration in tenebrionid beetles from the Namib desert. J Comp Physiol 155 B:155-162

Bauer AM, Russell AP (1991) Pedal specializations in dune-dwelling geckos. J Arid Environ 20: 43-62

Belk D, Cole GA (1975) Adaptational biology of desert temporary-pond inhabitants. In: Hadley NF (ed) Environmental physiology of desert organisms. Dowden, Hutchinson, Ross, Stroudsburg, Penn, pp 207-226

Bellairs A (1969) The life of reptiles Vol II. Weidenfeld, Nicolson, London, pp 283-590

Bellairs A d'A, Bryant SV (1985) Autotomy and regeneration in reptiles. In: Gans C, Billett F (eds) Biology of the Reptilia 15 (Development B). Academic Press, London, pp 301-410

Bennett AF (1972) The effect of activity on oxygen consumption, oxygen debt, and heart rate in the lizards *Varanus gouldii* and *Sauromalus hispidus*. J Comp Physiol 79:259-280

Bennett AF (1973a) Ventilation in two species of lizards during rest and activity. Comp Biochem Physiol 46A:653-671

Bennett AF (1973b) Blood physiology and oxygen transport during activity in two lizards, *Varanus gouldii* and *Sauromalus hispidus*. Comp Biochem Physiol 46A:673-690

Bennett AF (1980) The thermal dependence of lizard behaviour. Anim Behav 28:752-762

Bennett A (1982) The energetics of reptilian activity. In: Gans C, Pough FH (eds) Biology of the Reptilia, 13 (Physiology D). Physiological ecology. Academic Press, London, pp 155-199

Bennett AF, Dawson WR (1972) Aerobic and anaerobic metabolism during activity in the lizard *Dipsosaurus dorsalis*. J Comp Physiol 81:289-299

Bennett AF, Dawson WR (1976) Metabolism. In: Gans C, Dawson WR (eds) Biology of the Reptilia 5 (Physiology A). Academic Press, London, pp 127-223

Bennett AF, Gleeson TT (1976) Activity metabolism in the lizard *Sceloporus occidentalis*. Physiol Zoöl 49:65-76

Bennett AF, Gleeson TT (1979) Metabolic expenditure and the cost of foraging in the lizard *Cnemidophorus murinus*. Copeia 1979(4):573-577
Bennett AF, Nagy KA (1977) Energy expenditure of free-ranging lizards. Ecology 58:697-700
Bentley PJ (1976) Osmoregulation. In: Gans C, Dawson WR (eds) Biology of the Reptilia, 5 (Physiology A). Academic Press, London, pp 365-412
Bentley PJ, Blumer WFC (1962) Uptake of water by the lizard *Moloch horridus*. Nature 194:699-700
Berk ML, Heath JE (1975) An analysis of behavioural thermoregulation in the lizard *Dipsosaurus dorsalis*. J Therm Biol 1:15-22
Berry KH (1974) The ecology and social behaviour of the chuckwalla *Sauromalus obesus obesus* Baird. Univ Calif Publ Zool 101:1-60
Best TL, James HC (1985) Rattlesnakes (genus *Crotalus*) of the Pedro Amandariz lava field, New Mexico. Copeia 1985 (1):213-215
Best TL, James HC, Best FH (1983) Herpetofauna of the Pedro Amandariz lava field New Mexico. Tex J Sci 35:225-245
Beuchat CA (1986) Reproductive influences on the thermoregulatory behavior of a live-bearing lizard. Copeia 1986(4):971-979
Beuchat CA (1988) Temperature effects during gestation in a viviparous lizard. J Therm Biol 13:135-142
Beuchat CA, Ellner S (1987) A quantitative test of life history theory: thermoregulation by a viviparous lizard. Ecol Monogr 57:45-60
Beuchat CA, Vleck D, Braun EJ (1986) Role of the urinary bladder in osmatic regulation of neonatal lizards. Physiol Zool 59:539-551
Bjerke R, Lundheim R, Rein JO, Strømme JA, Zachariassen KE (1989) Water balance and osmotic regulation of the East African scorpion *Lychas burdoi* (Simon). Fauna Norv B36:33-36
Bligh J, Johnson KG (1973) A glossary of terms for thermal physiology. J Appl Physiol 35:941-961
Bodenheimer FS (1934) Studies on the ecology of Palestinian Coleoptera II. Seasonal and diurnal appearance and activity. Bull Soc R Entomol Egypte 18:211-241
Bodenheimer FS (1953) Problems on animal ecology and physiology in deserts. Spec Publ Res Counc Isr No 2:205-229
Bolwig N (1957) Experiments on the regulation of body temperature of certain tenebrionid beetles. J Entomol Soc S Afr 20:454-458
Bourne AR, Taylor JL, Watson TG (1986) Effect of temperature on the seasonal production of testicular androgens, in vitro by the lizard *Tiliqua rugosa*. Comp Biochem Physiol 85A:527-530
Boyer DR (1965) Ecology of the basking habitat in turtles. Ecology 46:99-118
Bradley RA (1986) The relationship between population density of *Paruroctonus utahensis* (Scorpionida: Vaejovidae) and characteristics of its habitat. J Arid Environ 11:165-171
Bradley RA (1988a) The behavioural ecology of scorpions — a review. In: Austin AD, Heather NW (eds) Australian arachnology. Aust Entomol Soc Misc Publ 5:23-36
Bradley RA (1988b) The influence of weather and biotic factors on the behaviour of the scorpion (*Paruroctonus utahensis*). J Anim Ecol 57:533-551
Bradshaw SD (1970) Seasonal changes in the water and electrolyte metabolism of *Amphibolurus* lizards in the field. Comp Biochem Physiol 36:689-718
Bradshaw SD (1986) Ecophysiology of desert reptiles. Academic Press, Sydney
Bradshaw SD (1988) Desert reptiles: a case of adaptation or pre-adaptation? J Arid Environ 14:155-174
Bradshaw SD, Main AR (1968) Behavioural attitudes and regulation of temperature in *Amphibolurus* lizards. J Zool (Lond) 154:193-221
Bradshaw SD, Gans C, Saint Girons H (1980) Behavioural thermoregulation in a pygopodid lizard *Lialis burtonis*. Copeia 1980(4):738-743
Brattstrom BH (1971) Critical thermal maxima of some Australian skinks. Copeia 1971(3):554-557
Braysher ML (1976) The excretion of hyperosmotic urine and other aspects of the electrolyte balance of the lizard *Amphibolurus maculosus*. Comp Biochem Physiol 54A:341-345
Brown JH (1986) The roles of vertebrates in desert ecosystems. pp 51-71 in Whitford (1986)
Brown JH, Reichman OJ, Davidson DW (1979) Granivory in desert ecosystems. Annu Rev Ecol Syst 10:201-227
Brown KM, Sexton OJ (1973) Stimulation of reproductive activity of female *Anolis sagrei* by moisture. Physiol Zoöl 46:168-172

Brownell PH (1977) Compressional and surface waves in sand: used by desert scorpions to locate prey. Science (NY) 197:479-482

Brownell PH, Farley RD (1979a) Prey-localizing behaviour of the nocturnal desert scorpion *Paruroctonus mesaensis*: orientation to substrate vibrations. Anim Behav 27:185-193

Brownell PH, Farley RD (1979b) Orientation to vibrations in sand by the nocturnal scorpion *Paruroctonus mesaensis*: mechanism of target localization. J Comp Physiol A131:31-38

Broza M (1979) Dew, fog and hygroscopic food as a source of water for desert arthropods. J Arid Environ 2:43-49

Broza M, Borut A, Pener MP (1976) Osmoregulation in the desert tenebrionid beetle *Trachyderma philistina* Reiche during dehydration and subsequent rehydration. Isr J Med Sci 12:868-871

Bucherl W, Buckley EE (eds) (1967-1971) Venomous animals and their venoms. Vol I, II Venomous vertebrates. Vol III Venomous invertebrates. Academic Press, New York

Burghardt GM (1977) Learning processes in reptiles. In: Gans C, Tinkle DW (eds) Biology of the Reptilia 7 (Ecology and behaviour A). Academic Press, London, pp 555-681

Burkholder GL, Walker JM (1973) Habitat and reproduction of the desert whiptail lizard, *Cnemidophorus tigris* Bird and Girard in southwestern Idaho at the northern part of its range. Herpetologica 29:76-83

Burrage BR (1973) Comparative ecology and behaviour of *Chamaeleo pumilus pumilus* (Gmelin) and *C. namaquensis* A. Smith (Sauria: Chamaeleontidae). Ann S Afr Mus 61:1-158

Bury RB, Balcooyen TG (1976) Temperature selectivity in the legless lizard, *Anniella pulchra*. Copeia 1976(1):152-155

Busack SD (1975) Biomass estimates and thermal environment of a population of the fringed-toed lizard, *Acanthodactylus pardalis*. Brit J Herpetol 5:457-459

Bustard HR (1967a) A mechanism for greater predator survival during cold torpor in gekkonid lizards. Brit J Herpetol 4:7-8

Bustard HR (1967b) Defensive display behaviour of the Australian gecko *Nephrurus asper*. Herpetologica 23:126-129

Bustard HR (1968a) Temperature dependent tail autotomy mechanism in gekkonid lizards. Herpetologica 24:127-130

Bustard HR (1968b) The ecology of the Australian gecko *Gehyra variegata* (Duméril and Bibron) in northern New South Wales. J Zool (Lond) 154:113-134

Bustard RH (1968c) *Pygopus nigriceps* (Fischer): a lizard mimicking a venomous snake. Br J Herpetol 4:22-24

Bustard RH (1979) Defensive mechanisms and predation in gekkonid lizards. Br J Herpetol 6:9-11

Büttiker W, Anderson E (1988) Water-extracting tenebrionid beetles from the Wahiba Sands, Oman. J Oman Stud Spec Rep No 3:321-324

Buxton PA (1923) Animal life in deserts. A study of the fauna in relation to the environment. Edward Arnold, London

Buxton PA (1924) Heat, moisture and animal life in deserts. Proc R Soc (B) 96:123-131

Campbell HW (1969) The effects of temperature on the auditory sensitivity of lizards. Physiol Zoöl 42:183-210

Cannon JG, Kluger MJ (1985) Altered thermoregulation in the iguana *Dipsosaurus dorsalis* following exercise. J Therm Biol 10:41-45

Carlisle DB (1968) Triops (Entomostraca) eggs killed only by boiling. Science (NY) 161:279-280

Carlisle DB, Ellis PE, Betts E (1965) The influence of aromatic shrubs in sexual maturation in the desert locust *Schistocerca gregaria*. J Insect Physiol 11:1541-1558

Carpenter CC (1978) Tongue display by the common bluetongue (*Tiliqua scincoides*) (Reptilia, Lacertilia, Scincidae). J Herpetol 12:428-429

Case T (1976) Seasonal aspects of thermoregulatory behavior in the chuckawalla, *Sauromalus obesus* (Reptilia, Lacertilia, Iguanidae). J Herpetol 10:85-95

Casey TH (1988) Thermoregulation and heat exchange. Adv Insect Physiol 20:119-146

Chapman BM, Chapman RF (1964) Observations on the biology of the lizard *Agama agama* in Ghana. Proc Zool Soc Lond 143:121-132

Chapman RF (1959) Field observations on the behaviour of hoppers of the red locust (*Nomadacris septemfasciata* Serville). Anti-Locust Bull 33:1-51

Chew RM (1961) Water metabolism of desert-inhibiting vertebrates. Biol Rev 36:1-31

Chew RM, De Vita J (1980) Foraging characteristics of a desert ant assemblage: functional morphology and species separation. J Arid Environ 3:75-83
Chong G, Heatwole H, Firth BT (1973) Panting thresholds of lizards — II Diel variation in the panting threshold of *Amphibolurus muricatus*. Comp Biochem Physiol 46A:827-829
Cloudsley-Thompson JL (1956) Studies in diurnal rhythms. VI Bioclimatic observations in Tunisia and their significances in relation to the physiology of the fauna, especially woodlice, centipedes, scorpions and beetles. Ann Mag Nat Hist (12)9:305-329
Cloudsley-Thompson JL (1958) Spiders, scorpions, centipedes and mites. Pergamon, London
Cloudsley-Thompson JL (1961a) Adaptive functions of circadian rhythms. Cold Spring Harb Symp Quant Biol (1960) 25:345-355
Cloudsley-Thompson JL (1961b) Rhythmic activity in animal physiology and behaviour. Academic Press, New York
Cloudsley-Thompson JL (1961c) Some aspects of the physiology and behaviour of *Galeodes arabs*. Entomol Exp Appl 4:257-263
Cloudsley-Thompson JL (1961d) Observations on the natural history of the 'camel spider' *Galeodes arabs* CL Koch (Solifugae: Galeodidae) in the Sudan. Entomol Mon Mag 97:145-152
Cloudsley-Thompson JL (1962a) Bioclimatic observations in the Red Sea hills and coastal plain, a major habitat of the desert locust. Proc R Ent Soc Lond (A)37:27-34
Cloudsley-Thompson JL (1962b) Some aspects of the physiology and behaviour of *Dinothrombium* (Acari). Entomol Exp Appl 5:69-73
Cloudsley-Thompson JL (1962c) Lethal temperatures of some desert arthropods and the mechanism of heat death. Entomol Exp Appl 5:270-280
Cloudsley-Thompson JL (1964a) Terrestrial animals in dry heat: arthropods. In: Dill DB (ed) Handbook of physiology, Sect 4. Am Physiol Soc, Washington DC, pp 451-465
Cloudsley-Thompson JL (1964b) On the function of the sub-elytral cavity in desert Tenebrionidae. Entomol Mon Mag 100:148-151
Cloudsley-Thompson JL (1965a) Rhythmic activity, temperature-tolerance, water-relations and mechanism of heat death in a tropical skink and gecko. J Zool (Lond) 146:55-69
Cloudsley-Thompson JL (1965b) The lethal temperature of *Triops granarius* (Lucas) (Branchiopoda: Notostraca). Hydrobiologia 25:424-425
Cloudsley-Thompson JL (1965c) Desert life. Pergamon, Oxford
Cloudsley-Thompson JL (1965d) The scorpion. Sci J 1:35-41
Cloudsley-Thompson JL (1966) Orientation responses of *Triops granarius* (Lucas) (Branchiopoda: Notostraca) and *Streptocephalus* spp. (Branchiopoda:Anostraca). Hydrobiologia 27:33-38
Cloudsley-Thompson JL (1968) The Merkhiyat jebels: a desert community. In: Brown GW Jr (ed) Desert biology, Vol 1. Academic Press, New York, pp 1-20
Cloudsley-Thompson JL (1969a) The zoology of tropical Africa. Weidenfeld & Nicolson, London. Norton, New York (The World Naturalist)
Cloudsley-Thompson JL (1969b) Acclimation, water and temperature relations of the woodlice *Metoponorthus pruinosus* and *Perischyphis jannonei* in the Sudan. J Zool (Lond) 158:267-276
Cloudsley-Thompson JL (1970a) On the biology of the desert tortoise *Testudo sulcata* in Sudan. J Zool (Lond) 160:17-33
Cloudsley-Thompson JL (1970b) Terrestrial invertebrates. In: Whittow GC (ed) Comparative physiology of thermoregulation, Vol 1. Academic Press, London, pp 15-77
Cloudsley-Thompson JL (1971) The temperature and water relations of reptiles. Merrow, Watford Herts
Cloudsley-Thompson JL (1972a) Temperature regulation in desert reptiles. In: Maloiy GMO (ed) Comparative physiology of desert animals. Symp Zool Soc Lond No 31:39-59
Cloudsley-Thompson JL (1972b) Site tenure and selection in the African gecko *Tarentola annularis* (Geoffroy). Br J Herpetol 4:286:292
Cloudsley-Thompson JL (1973) Factors influencing the super-cooling of tropical Arthropoda, especially locusts. J Nat Hist 7:471-480
Cloudsley-Thompson JL (1974) Physiological thermoregulation in the spurred toroise (*Testudo graeca* L.). J Nat Hist 8:577:587
Cloudsley-Thompson JL (1975) Adaptations of Arthropoda to arid environments. Annu Rev Entomol 20:261-283

Cloudsley-Thompson JL (1977a) Man and the biology of arid zones. Edward Arnold, London
Cloudsley-Thompson JL (1977b) Adaptational biology of Solifugae (Solpugida). Bull Brit Arachnol Soc 4:61-71
Cloudsley-Thompson JL (1977c) The water and temperature relations of woodlice. Meadowfield Press, Shildon Co Durham (Patterns of Progress Zoology, Vol 8)
Cloudsley-Thompson JL (1977d) Some aspects of the biology of *Amphisbaena alba*. Br J Herpetol 5:617-621
Cloudsley-Thompson JL (1977e) Man and the biology of arid zones. Edward Arnold, London (Contemporary Biology)
Cloudsley-Thompson JL (1978) Biological clocks in Arachnida. Bull Br Arachnol Soc 4:61-71
Cloudsley-Thompson JL (1979a) Adaptive functions of the colours of desert animals. J Arid Environ 2:95-104
Cloudsley-Thompson JL (1979b) Water loss and locomotory activity in *Agama persica* and *Diplometopon zarudnyi* from Kuwait. J Arid Environ 2:273-277
Cloudsley-Thompson JL (1980a) Biological clocks. Their functions in nature. Weidenfeld, Nicolson, London
Cloudsley-Thompson JL (1980b) Tooth and claw. Defensive strategies in the animal world. JH Dent, London
Cloudsley-Thompson JL (1981a) Bionomics of the rainbow lizard *Agama agama* (L.) in eastern Nigeria during the dry season. J Arid Environ 4:235-245
Cloudsley-Thompson JL (1981b) A comparison of rhythmic locomotory activity in tropical forest Arthropoda with that in desert species. J Arid Environ 4:327-334
Cloudsley-Thompson JL (1983) Desert adaptations in spiders. J Arid Environ 6:307-317
Cloudsley-Thompson JL (1986) Desert adaptations in arachnids. Proc 9th Int Congr Arachnol, Panama 1983. Smithsonian Institution Press, Washington DC, pp 29-32
Cloudsley-Thompson JL (1988a) Evolution and adaptation of terrestrial arthropods. Springer, Berlin Heidelberg New York Tokyo
Cloudsley-Thompson JL (1988b) Strategies for survival of desert reptiles. Sci Rev Arid Zone Res 6:187-288
Cloudsley-Thompson JL (1989a) Temperature and the activity of ants and other insects in Central Australia. J Arid Environ 16:185-192
Cloudsley-Thompson JL (1989b) Behavioural adaptation of desert Arthropoda. Boll Acc Gioenia Sci Nat (1987) 20(332):283-293
Cloudsley-Thompson JL (1990) Thermal ecology and behaviour of *Physadesmia globosa* (Coleoptera:Tenebrionidae) in the Namib Desert. J Arid Environ 19:317-324
Cloudsley-Thompson JL, Chadwick MJ (1964) Life in deserts. Foulis, London
Cloudsley-Thompson JL, Constantinou C (1984) Stridulatory apparatus of Solifugae [Solpugida]. J Arid Environ 7:365:369
Cloudsley-Thompson JL, Crawford CS (1970a) Lethal temperatures of some arthropods of the southwestern United States. Entomol Mon Mag 106:26-29
Cloudsley-Thompson JL, Crawford CS (1970b) Water and temperature relations, and diurnal rhythms of scolopendromorph centipedes. Entomol Exp Appl 13:187-193
Cloudsley-Thompson JL, Idris BM (1964) The insect fauna of the desert near Khartoum: seasonal fluctuations and the effect of grazing. Proc R Entomol Soc Lond (A)39:41-46
Cochran DG (1985) Nitrogenous excretion. In: Kerkut GA, Gilbert LI (eds) Comprehensive insect biochemistry and pharmacy, Vol 4. Pergamon, Oxford, pp 467-506
Coenen-Stass D (1989) Investigations on microclimatic and climatic preference of the desert woodlouse *Hemilepistus reaumuri* (Audouin & Savigny, 1826) (Oniscidea Isopoda Crustacea). Monit Zool Ital (N.S.) Monogr 4:425-434
Coineau Y, Lancaster N, Prodon R, Seely MK (1982) Burrowing habits and substrate selection in ultrapsammophilous tenebrionid beetles of the Namib desert. Vie Milieu 32:125-131
Collins MS, Haverty MI, La Fage JP, Nutting WL (1973) High-temperature tolerance in two species of subterranean termites from the Sonoran Desert in Arizona. Environ Entomol 2:1122-1123
Congdon JD, Vitt LH, King WW (1974) Geckos: adaptive significance and energetics of tail autotomy. Science (NY) 184:1379-1380

Constantinou C (1980) Entrainment of the circadian rhythm of activity in desert and forest inhabiting scorpions. J Arid Environ 3:133-139
Constantinou C, Cloudsley-Thompson JL (1984) Stridulatory structures in scorpions of the families Scorpionidae and Diplocentridae. J Arid Environ 7:359-364
Constantinou C, Cloudsley-Thompson JL (1985) The circadian rhythm of locomotory activity in the desert lizard *Acanthodactylus schmidti*. J Interdiscip Cycle Res 16:107-111
Constantinou C, Cloudsley-Thompson JL (1987) Water relations of some woodlice (Oniscidea) from arid and mesic environments. Qatar Univ Sci Bull 7:203-216
Cooke JAL, Roth VD, Miller FH (1972) The urticating hairs of theraphosid spiders. Am Mus Novit No 2498:1-43
Cooper PD (1982) Water balance and osmoregulation in a free-ranging tenebrionid beetle, *Onymacris unguicularis*, of the Namib Desert. J Insect Physiol 28:737-742
Cott HB (1940) Adaptive coloration in animals. Methuen, London
Coutchié PA, Crowe JH (1979) Transport of water vapor by tenebrionid beetles. I Kinetics. Physiol Zool 52:67-87
Cowles RB (1958) Possible origin of dermal temperature regulation. Evolution 12:347-357
Cowles RB, Bogert CM (1944) A preliminary study of thermal requirements of desert reptiles. Bull Am Mus Nat Hist 83:265-296
Crawford CS (1972) Water relations in a desert millipede *Orthoporus ornatus* (Girard) (Spirostreptidae). Comp Biochem Physiol 42A:521-535
Crawford CS (1978) Seasonal water balance in *Orthoporus ornatus*, a desert millipede. Ecology 59:996-1004
Crawford CS (1979) Desert detritivores: a review of life history patterns and trophic roles. J Arid Environ 2:31-42
Crawford CS (1981) Biology of desert invertebrates. Springer, Berlin Heidelberg New York
Crawford CS (1986) The role of invertebrates in desert ecosystems. pp 51-91 In Whitford (1986)
Crawford CS, Cloudsley-Thompson JL (1971) Water relations and desiccation-avoiding behaviour in the vinegaroon *Mastigoproctus giganteus* (Arachnida: Uropygi). Entomol Exp Appl 14:99-106
Crawford CS, Matlack MC (1979) Water relations of desert millipede larvae, larva-containing pellets, and surrounding soil. Pedobiologia 19:48-55
Crawford CS, McClain EL (1983) Habitat use during winter relative to water balance in a harpagophorid millipede from the Namib Desert. Madoqua 13:243-250
Crawford CS, Riddle WA (1975) Overwintering physiology of the scorpion *Diplocentrus spitzeri*. Physiol Zoöl 48:84-92
Crawford CS, Seely MK (1987) Assemblages of surface-active arthropods in the Namib dunefield and associated habitats. Rev Zool Afr 101:397-421
Crawford CS, Taylor EC (1984) Decomposition in arid environments: role of the detritivore gut. S Afr J Sci 80:170-176
Crawford CS, Warburg MR (1982) Water balance and apparent oocyte resorption in desert millipedes. J Exp Zool 222:215-226
Crawford CS, Wooten RC Jr (1973) Water relations in *Diplocentrus spitzeri*, a semimontane scorpion from the southwestern United States. Physiol Zoöl 46:218-229
Crawford CS, Riddle WA, Pugach S (1975) Overwintering physiology of the centipede *Scolopendra polymorpha*. Physiol Zoöl 48:290-294
Crawford CS, Goldenberg S, Warburg MR (1986) Seasonal water balance in *Archispirostreptus syriacus* (Diplopoda:Spirostreptidae) from mesic and xeric Mediterranean environments. J Arid Environ 10:127-136
Crawford CS, Hanrahan SA, Seely MK (1991) Scale-related habitat use of *Physadesmia globosa* (Coleoptera: Tenebrionidae) in a riparian desert environment. In: Seely MK (ed) Namib ecology: 25 years of Namib research. Transvaal Mus Monogr No 7:135-142
Creusere FM, Whitford WG (1981) Temporal and spatial resource partitioning in a Chihuahuan desert lizard community. In:Scott NJ (ed) Herpetological communities. US Department of the Interior, Fish and Wildlife Service Wildlife, Wildlife Research Report 13:121-127
Crisp M, Cook LM, Hereward FV (1979) Colour and heat balance in the lizard *Lacerta dugesii*. Copeia 1979(2):250-258

Crovetti A (1970) Note eco-etologiche sulla entomofauna primaverile dello 'Uádi Caám' (Tripolitania). Resultati delle missioni entomolgiche dei proff G Fiori ed E Mellini nel Nord-Africa XXIV. Studi Sassaresi 111 Ann Fac Agr Univ Sassari 18:1-114

Crowley SR (1987) The effect of desiccation upon the preferred body temperature and activity level of the lizard *Sceloporus undulatus*. Copeia 1987(1):25-32

Crowley SR, Pietruszka RD (1983) Aggressiveness and vocalization in the leopard lizard (*Gambelia wislizenii*). Anim Behav 31:1055-1066

Cuellar HS, Cuellar O (1977) Evidence for endogenous rhythmicity in the reproductive cycle of the parthenogenetic lizard *Cnemidophorus uniparens* (Reptilia:Teiidae). Copeia 1977(3):554-557

Curry-Lindahl K (1979) Thermal ecology of the tree agama (*Agama atricollis*) in Zaire with a review of heat tolerance in reptiles. J Zool (Lond) 188:185-220

Curtis BA (1985a) Temperature tolerances in the Namib Desert dune ant, *Camponotus detritus*. J Insect Physiol 31:463:466

Curtis BA (1985b) Observations on the natural history and behaviour of the dune ant, *Camponotus detritus* Emery, in the central Namib Desert. Madoqua 14:279-289

Dantzler WH (1976) Renal function (with special emphasis on nitrogen excretion). In: Gans C, Dawson WR (eds) Biology of the Reptilia 5 (Physiology A). Academic Press, London, pp 447-503

Dantzler WH, Holmes WN (1974) Water and mineral metabolism in Reptilia. Chemical Zoology 9:277-336. Academic Press, New York

Darlington JPEC (1982) The underground passages and storage pits used in foraging by a nest of the termite *Macrotermes michaelseni* in Kajiado, Kenya. J Zool (Lond) 198:237-247

Dawson WR (1967) Interspecific variation in physiological responses of lizards to temperature. In: Milstead WW (ed) Lizard ecology: a symposium. Univ Missouri Press, Columbia, pp 230-257

Dawson WR (1975) On the physiological significance of the preferred body temperatures of reptiles. In: Gates DM, Schmerl R (eds) Perspectives of biophysical ecology. Springer, Berlin Heidelberg New York, pp 443-473

Dawson WR, Shoemaker VH, Licht P (1966) Evaporative water losses of some small Australian lizards. Ecology 47:589-594

Délye G (1968) Recherches sur l'écologie, la physiologie et l'éthologie des fourmis du Sahara. Thèse, Université d'Aix-Marseille, Marseille

Denlinger DL (1986) Dormancy in tropical insects. Annu Rev Entomol 31:239-264

De Witt CB (1967) Behavioral thermoregulation in the desert iguana. Science (NY) 158:809-810

Dial BE (1978a) The thermal ecology of two sympatric, nocturnal *Coleonyx* (Lacertilia:Gekkonidae). Herpetologica 34:194-201

Dial BE (1978b) Aspects of the behavioral ecology of two Chihuahuan desert geckos (Reptilia, Lacertilia, Gekkonidae). J Herpetol 12:209-216

Dmi'el R, Borut A (1972) Thermal behavior, heat-exchange, and metabolism in the desert snake *Spalerosophis cliffordi*. Physiol Zoöl 45:78-94

Dmi'el R, Zilber B (1971) Water balance in a desert snake. Copeia 1971(4):754-755

Dreisig H (1980) Daily activity, thermoregulation and water loss in the tiger beetle, *Cicindela hybrida*. Oecologia (Berl) 44:376-389

Dreisig H (1985) A time budget model of thermoregulation in shuttling ecotherms. J Arid Environ 8:191-205

Dunham AE (1978) Food availability as a proximate factor influencing individual growth rates in the iguanid lizard *Sceloporus merriami*. Ecology 59:770-778

Duvall D, Guillette LJ Jr, Jones RE (1982) Environmental control of reptilian reproductive cycles. In: Gans C, Pough FH (eds) Biology of the Reptilia 13 (Physiology D). Physiological ecology. Academic Press, London, pp 201-231

Duvdevani I, Borut A (1974) Mean body temperature and heat absorption in four species of *Acanthodactylus* lizards (Lacertidae). Herpetologica 30:176-181

Eberling L, Paarmann W (1985) Diel activity patterns of the desert carabid beetle *Thermophilum* (= *Anthia*) *sexmaculata* F. (Coleoptera:Carabidae). J Arid Environ 8:141-155

Edmunds M (1974) Defence in animals. A survey of anti-predator defences. Longman, Harlow Essex

Edney EB (1966a) Animals of the desert. In: Hills ES (ed) Arid lands. Methuen, London, pp 181-218

Edney EB (1966b) Absorption of water vapour from unsaturated air by *Arenivaga* sp. (Polyphagidae, Dictyoptera). Comp Biochem Physiol 19:387-408

Edney EB (1968) The effect of water loss on the haemolymph of *Arenivaga* sp. and *Periplaneta americana*. Comp Biochem Physiol 25:149-158

Edney EB (1971a) The body temperature of tenebrionid beetles in the Namib desert of southern Africa. J Exp Biol 55:253-272

Edney EB (1971b) Some aspects of water balance in tenebrionid beetles and a thysanuran from the Namib Desert of southern Africa. Physiol Zoöl 44:61-76

Edney EB (1974) Desert arthropods. In: Brown GW (ed) Desert biology, Vol II. Academic Press, New York, pp 311-384

Edney EB (1977) Water balance in land arthropods. Springer, Berlin Heidelberg New York

Edney EB, Nagy KA (1976) Water balance and excretion. In: Bligh J, Cloudsley-Thompson JL, Macdonald AG (eds) Environmental physiology of animals. Blackwell, Oxford. Wiley, New York, pp 106-132

Edney EB, Haynes S, Gibo D (1974) Distribution and activity of the desert cockroach *Arenivaga investigata* (Polyphagidae) in relation to microclimate. Ecology 55:420-427

Edney EB, Franco P, Wood R (1978) The responses of *Arenivaga investigata* (Dictyoptera) to gradients of temperature and humidity in sand studied by tagging with technetium 99M. Physiol Zoöl 51:241-255

Edwards HT, Dill DB (1935) Properties of reptilian blood. II The Gila monster (*Heloderma suspectum* Cope). J Cell Comp Physiol 6:21-35

Ellis PE, Ashall C (1957) Field studies on diurnal behaviour, movement and aggregation in the desert locust (*Schistocerca gregaria*) Forskal. Anti-locust Bull 25:1-94

El Rayah EA (1970a) Some reactions of two desert beetles, *Adesmia antiqua* and *Pimelia grandis* (Tenebrionidae) to temperature. Entomol Exp Appl 13:286-292

El Rayah EA (1970b) Humidity responses of two desert beetles, *Adesmia antiqua* and *Pimelia grandis* (Coleoptera:Tenebrionidae). Entomol Exp Appl 13:438-447

Engbretson GA, Hutchson VH (1976) Parietalectomy and thermal selection in the lizard *Sceloporus magister*. J Exp Zool 198:29-38

Engbretson GA, Livezey RL (1972) The effects of aggressive display on body temperature in the fence lizard *Sceloporus occidentalis occidentalis* Baird and Girard. Physiol Zoöl 45:247-254

Erberling L, Paarmann (1985) Diel activity patterns of the desert carabid beetle *Thermophilum* (= *Anthia*) *sexmaculatum* F. (Coleoptera:Carabidae). J Arid Environ 8:141-155

Fiori G (1956) Ecologia ed etologia della entomofauna dello 'Uadi Caam'. Boll Inst Entomol Univ Bologna 22:1-44

Fiori G (1977) La cavità sottoelitrale die tenebrionidi apomorfi. Redia 60:1-111

Firth BT, Heatwole H (1976) Panting thresholds of lizards: the role of the pineal complex in panting processes in an agamid, *Amphibolurus muricatus*. Gen Comp Endocrinol 29:388-401

Firth BT, Webb GJW, Johnson CR (1972) Effect of time of day and photoperiod on heart rate in the scincid lizard, *Tiliqua scincoides*. Comp Biochem Physiol 43A:805-808

Fleissner G, Fleissner G (1985) Neurobiology of a circardian clock in the visual system of scorpions. In: Barth FD (ed) Neurobiology of arachnids. Springer, Berlin Heidelberg New York Tokyo, pp 351–375

Fogleman JC, Heed WP (1989) Column cacti and desert *Dropsophila*: the chemistry of host plant specificity. pp 1-24 in Schmidt (1989)

Formanowicz DR Jr, Bradley PJ (1987) Fluctuations in prey density: effects on the foraging tactics of scolopendrid centipedes. Anim Behav 35:453-461

Fox SF (1978) Natural selection on behavioral phenotypes of the lizard *Uta stansburiana*. Ecology 59:834-847

Fox SF, Rose E, Myers R (1981) Dominance and the acquisition of superior home ranges in the lizard *Uta stansburiana*. Ecology 62:888-893

Fox SF, Heger NA, Delay LS (1990) Social cost of tail loss in *Uta stansburiana*: lizard tails as status — signalling badges. Anim Behav 39:549-554

Frankenberg E (1976) Daily activity of geographical forms of *Ptyodactylus* (Reptilia:Gekkoninae). Isr J Zool 25:207

Frankenberg E (1978) Interspecific and seasonal variation of daily activity times in gekkonid lizards (Reptilia, Lacertilia). J Herpetol 12:505-519

Frankenberg E (1979) Influence of light and temperature on daily activity patterns of three Israeli forms of *Ptyodactylus* (Reptilia:Gekkonidae). J Zool (Lond) 189:21-30

Frankenberg E (1981) Optomotor responses of three congeneric gekkonid lizards having different daily activity times. J Zool (Lond) 193:147-156

Fraser SP (1985) Variability of hearing and cooling rates during radiant healing in a scincid lizard, *Egernia cunninghami*. Comp Biochem Physiol 80A:281-286

Fusari MH (1985) Drinking of soil water by the California legless lizard *Anniella pulchra*. Copeia 1985(4):981-986

Gans C (1974) Biomechanics: an approach to vertebrate biology. JB Lippincott, Philadelphia

Gans C, Dawson RW (1976) Reptilia physiology: an overview. In: Gans C, Dawson RW (eds) Biology of the Reptilia 5 (Physiology A). Academic Press, London, pp 1-17

Gans C, Huey RB (eds) (1988) Biology of the Reptilia 16 (Ecology B). Defense and life history. Alan Liss, New York

Gans C, Maderson PFA (1973) Sound producing mechanisms in recent reptiles: review and comment. Am Zool 13:1195-1203

Gans C, Krakauer T, Paganell CB (1968) Water loss in snakes: interspecific and intraspecific variability. Comp Biochem Physiol 27:747-761

Gans C, Merlin R, Blumer WFC (1982) The water-collecting mechanism of *Moloch horridus* re-examined. Amphib Reptilia 3:57-64

Garland T Jr (1985) Ontogenetic and individual variation in size, shape and speed in the Australian agamid lizard *Amphibolurus nuchalis*. J Zool (Lond) 207:425-439

Gatten RE (1974) Effect of nutritional status of the preferred body temperature of the turtles *Psudemys scripta* and *Terrapene ornata*. Copeia 1944(4):912-917

George W (1986) The thermal niche: desert sand and desert rock. J Arid Environ 10:213-224

Gersani M, Degen AA (1988) Daily energy intake and expenditure of the weaver ant *Polyrhachis simplex* (Hymenoptera:Formicidae) collecting honey dew from the cicada *Oxyrrhachis versicolor* (Homoptera:Membracidae). J Arid Environ 15:75-80

Gilles-Baillien M (1974) Seasonal variations in reptiles. In: Florkin M, Scheer BT (eds) Chemical Zoology 9 (Amphibia and Reptilia). Academic Press, New York, pp 353-377

Goldberg SR, Rodriguez E (1986) Reproductive cycles of two iguanid lizards from northern Chile, *Tropidurus quadrivittatus* and *Tropidurus theresioides*. J Arid Environ 10:147-151

Goodall DW, Perry RA, Howes KMW (eds) (1981) Arid-land ecosystems: structure, functioning and management Vol 2. Cambridge University Press, Cambridge

Goode MJ, Duvall D (1989) Body temperature and defensive behaviour of free-ranging prairie rattlesnakes *Crotalus viridis viridis*. Anim Behav 38:360-362

Gordon DM (1983) The relation of recruitment rate to activity rhythms in the harvester ant, *Pogonomyrmex barbatus* (F Smith) (Hymenoptera:Formicidae). J Kansas Entomol Soc 56:277-285

Gordon DM (1987) Group-level dynamics in harvester ants: young colonies and the role of patrolling. Anim Behav 35:833-843

Gordon DM (1988) Nest-plugging: interference and competition in desert ants (*Novomessor cockerelli* and *Pogonomyrmex barbatus*). Oecologia (Berl) 75:114-118

Gordon DM (1989) Dynamics of task switching in harvester ants. Anim Behav 38:194-204

Grant BW, Dunham AE (1988) Thermally imposed time constraints on the activity of the desert lizard *Sceloporus merriami*. Ecology 69:167-176

Grant C (1957) Autophagy in *Eumeces*. Herpetologica 13:156

Graves B (1979) Defensive behavior of female prairie rattlesnakes (*Crotalus viridis*) changes after parturition. Copeia 1989(3):791-794

Green B, King D (1978) Home range and activity patterns of the sand goanna, *Varanus gouldii* (Reptilia:Varanidae). Aust J Wildl Res 5:417-424

Green E, Mendelshson H (1989) Activity patterns and thermoregulatory behavior of the Egyptian tortoise *Testudo kleinmanni* in Israel. J Herpetol 23:404-409

Greenaway P (1981) Temperature limits to trailing activity in the Australian arid-zone meat ant *Iridomyrmex purpureus* form *viridiaeneus*. Aust J Zool 29:621-630

Greenberg N (1977) An ethogram of the blue spiny lizard, *Sceloporus cyanogenys* (Reptilia Lacertilia Iguanidae). J Herpetol 11:177-195

Greene HW (1988) Antipredator mechanisms in reptiles. In: Gans C, Huey RB (eds) Biology of the Reptilia 16 (Ecology B). Defense and life history. Alan R Liss, New York, pp 1-152

Greer AE (1980) Critical thermal maximum temperatures in Australian scincid lizards: their ecology and evolutionary significance. Aust J Zool 28.91-102

Gregory PT (1982) Reptilian hibernation. In: Gans C, Pough FH (eds) Biology of the Reptilia 13 (Physiology D). Physiological ecology. Academic Press, London, pp 53-154

Griffiths RA (1984) The influence of light and temperature on diel activity rhythms in the sand boa *Eryx conicus*. J Herpetol 18:374-380

Grigg GC (1973) Some consequences of the shape and orientation of magnetic termite mounds. Aust J Zool 25:87-94

Grigg G, Jacklyn P, Taplan L (1988) The effects of buried magnets on colonies of *Amitermes* spp. building magnetic mounds in northern Australia. Physiol Entomol 13:285-289

Grubb P (1971) Comparative notes on the behavior of *Geochelone sulcata*. Herpetologica 27:328-333

Hadley NF (1970a) Micrometeorology and energy exchange in two desert arthropods. Ecology 51:434-444

Hadley NF (1970b) Water relations of the desert scorpion *Hadrurus arizonensis*. J Exp Biol 53:547-558

Hadley NF (1972) Desert species and adaptations. Am Sci 60:338-347

Hadley NF (1974) Adaptational biology of desert scorpions. J Arachnol 2:11-23

Hadley NF (1977) Epiculticular lipids of the desert tenebrionid beetle, *Eleodes armata*: seasonal and acclimatory effects on composition. Insect Biochem 7:277-283

Hadley NF (1979a) Recent developments in ecophysiological research on desert arthropods. J Arid Environ 2:211-218

Hadley NF (1979b) Thermal and water relations of desert animals. In: Perry RA, Goodall DW (eds) Arid-land ecosystems: structure functioning and management, Vol I. Cambridge Univ Press, Cambridge, pp 743-768 (International Biological Programme 6)

Hadley NF (1979c) Wax secretion and color phases of the desert tenebrionid beetle *Cryptoglossa verrucosa* (Le Conte). Science (NY) 203:367-369

Hadley NF (1981) Cuticular lipids of terrestrial plants and arthropods: a comparison of their structure, composition and waterproofing function. Biol Rev 56:23-47

Hadley NF (1982) Animal adaptations. In: Bender GL (ed) Reference handbook of the deserts of North America. Greenwood Press, Westport Connecticut, pp 405-418

Hadley NF (1984) Cuticle: ecological significance. In: Bereiter-Hahn J, Matoltsy AG, Richards KS (eds) Biology of the integument. Vol 1 Invertebrates. Springer, Berlin Heidelberg New York Tokyo, pp 685-693

Hadley NF (1985) The adaptive role of lipids in biological systems. John Wiley, New York

Hadley NF (1989) Lipid water barriers in biological systems. Prog Lipid Res 28:1-33

Hadley NF (1990) Environmental physiology. pp 321-340 in Polis (1990a)

Hadley NF, Hill RD (1969) Oxygen consumption of the scorpion *Centruriodes sculpturatus*. Comp Biochem Physiol 29:217-226

Hadley NF, Toolson EC, Quinlan MC (1989) Regional differences in cuticular permeability in the desert cicada *Diceroprocta apache*: implications for evaporative cooling. J Exp Biol 141:219-230

Hadley NF, Knisley CB, Schultz TD, Pearson DL (1990) Water relations of tiger beetle larvae (*Cicindela marutha*): correlations with habitat microclimate and burrowing activity. J Arid Environ 18:189-197

Hafez M, Ibrahim MM (1964) On the ecology and biology of the desert grasshopper, *Sphingonotus carenatus* Sauss., in Egypt. Bull Soc Entomol Egypte 48:193-217

Hamilton WJ III (1971) Competition and thermoregulatory behavior of the Namib Desert tenebrionid genus *Cardiosus*. Ecology 52:810-822

Hamilton WJ III (1973) Life's color code. McGraw-Hill, New York

Hamilton WJ III (1975) Coloration and its thermal consequences for diurnal desert insects. In: Hadley NF (ed) Environmental physiology of desert organisms. Dowden, Hutchinson, Ross, Stroudsburg Penn, pp 67-89

Hamilton WJ III, Coetzee CG (1969) Thermoregulatory behaviour of the vegetarian lizard *Angolosaurus skoogi* on the vegetationless northern Namib Desert dunes. Sci Pap Namib Desert Res Sta No 47:95-103

Hamilton WJ III, Seely MK (1976) Fog basking by the Namib Desert beetle *Onymacris unguicularis*. Nature 262:284-285

Hammerson GA (1979) Thermal ecology of the striped racer *Masticophis lateralis*. Herpetologica 35:267-273

Hanrahan SA, McClain E, Gernecke D (1984) Dermal glands concerned with the production of wax blooms in desert tenebrionid beetles. S Afr J Sci 80:176-181

Harlow P, Grigg G (1984) Shivering thermogenesis in a brooding diamond python *Python spilotes spilotes*. Copeia 1984(4):959-965

Hartline PH (1971) Physiological basis for detection of sound and vibration in snakes. J Exp Biol 54:349-371

Hauffe JC, Pietruezka RD, Seely MK (1988) Observations on the behaviour of *Onymacris laeviceps* Gebien (Coleoptera:Tenebrionidae:Adesmiini) in the central Namib Desert dunes. J Entomol Soc S South Afr 51:183-192

Hawke SD, Farley RD (1973) Ecology and behaviour of the desert burrowing cockroach *Arenivaga* sp. (Dictyoptera, Polyphagidae). Oecologia (Berl) 11:263-279

Heath JE (1965) Temperature regulation and diurnal activity in horned lizards. Univ Calif (Berkeley) Publ Zool 64:97-136

Heath JE (1966) Venous shunts in the cephalic sinuses of horned lizards. Physiol Zoöl 39:30-35

Heathcote RL (1983) The arid lands: their use and abuse. Longman, London

Heatwole H (1970) Thermal ecology of the desert dragon *Amphibolurus inermis*. Ecol Monogr 40:425-457

Heatwole H (1976) Reptile ecology. University of Queensland Press, St Lucia

Heatwole H (1977) Habitat selection in reptiles. In: Gans C, Tinkle DW (eds) Biology of the Reptilia 7 (Ecology and behaviour A). Academic Press, London, pp 137-155

Heatwole H, Butler H (1981) Structure of an assemblage of lizards on Barrow Island, Western Australia. Aust J Herpetol 1:37-44

Heatwole H, Firth BT (1982) Voluntary maximum temperature of the jackie lizard *Amphibolurus muricatus*. Copeia 1982(4):824-829

Heatwole H, Harrington S (1989) Heat tolerance of some ants and beetles from the pre-Saharan steppe of Tunisia. J Arid Environ 16:69-77

Heatwole H, Muir R (1979) Thermal microclimates in the pre-Saharan steppe of Tunisia. J Arid Environ 1:35-40

Heatwole H, Muir R (1989) Seasonal and daily activity of ants in the pre-Saharan steppe of Tunisia. J Arid Environ 16:49-67

Heinrich B (1975) Thermoregulation and flight energetics of desert insects. In: Hadley NF (ed) Environmental physiology of desert organisms. Dowden, Hutchinson, Ross, Stroudsburg Penn, pp 90-105

Heinrich B (1977) Why have some animals evolved to regulate a high body temperature? Am Nat 111:623-640

Henschel JR (1990) Spiders wheel to escape. S Afr J Sci 86:152

Henschel JR (1991) The biology of *Leucorchestris arenicola* (Araneae:Heteropodidae), a burrowing spider of the Namib dunes. In: Seely MK (ed) Namib ecology: 25 years of Namib research. Transvaal Mus Monogr No 7:115-127

Henwood K (1975) A field-tested thermoregulation model for two diurnal Namib Desert tenebrionid beetles. Ecology 56:1329-1342

Henzell RP (1982) Adaptation to aridity in lizards of the *Egernia whitei* species-group. In: Barker WR, Greenslade PJM (eds) Evolution of the flora and fauna of arid Australia. Peacock Publications, Frewville South Australia, pp 229-233

Hertz PE, Huey RB, Nevo E (1982) Fight versus flight: body temperature influences defensive responses of lizards. Anim Behav 30:676-679

Herzog HA Jr, Drummond H (1984) Tail autotomy inhibits tonic immobility in geckos. Copeia 1984(3):763-764

Hetherington TE (1989) Use of vibratory cues for detection of insect prey by the sandswimming lizard *Scincus scincus*. Anim Behav 37:290-297

Hillman S, Gorman GC, Thomas R (1979) Water loss in *Anolis* lizards: evidence for acclimation and intraspecific differences along a habitat gradient. Comp Biochem Physiol 62A:491-494

Hillyard SD, Vinegar A (1972) Respiration and thermal tolerance of the phyllopod crustacea *Triops longicaudatus* and *Thamnocephalus platyurus* inhabiting desert ephemeral ponds. Physiol Zoöl 45:189-195

Hinton HE (1960) Cryptobiosis in the larva of *Polypedilum vanderplanki* Hint. (Chironomidae). J Insect Physiol 5:286-300

Hinton HE (1981) Biology of insect eggs, Vol 1. Pergamon, Oxford

Hoffmann C (1967) Bau und Funktion der Trichobothrien von *Euscorpius carpathicus* L. Z Vergl Physiol 54:290-352

Hoffmann G (1984) Orientation behaviour of the desert woodlouse *Hemilepistus reaumuri*: adaptations to ecological and physiological problems. Symp Zool Soc Lond 53:405-422

Holm E (1973) The influence of constant temperatures upon the circadian rhythms of the Namib Desert dune lizard *Aporosaura anchiete* Bocage. Madoqua (II) 2:33-41

Holm E, Edney EB (1973) Daily activity in Namib Desert arthropods in relation to climate. Ecology 54:45-56

Holm E, Kirsten JF (1979) Pre-adaptation and speed mimicry among Namib Desert scarabaeids with orange elytra. J Arid Environ 2:263-271

Holm E, Scholtz CH (1980) Structure and pattern of the Namib Desert dune ecosystem at Gobabeb. Madoqua 12:3-39

Huey RB (1974) Winter thermal ecology of the iguanid lizard *Tropidurus peruvianus*. Copeia 1974(1):149-155

Huey RB (1982) Temperature physiology and the ecology of reptiles. In: Gans C, Pough FH (eds) Biology of the Reptilia 12 (Physiology C). Physiological ecology. Academic Press, London, pp 25-91

Huey RB, Pianka ER (1974) Ecological character displacement in a lizard. Am Zool 14:1127-1136

Huey RB, Pianka ER (1977a) Natural selection for juvenile lizards mimicking noxious beetles. Science (NY) 195:201-203

Huey RB, Pianka ER (1977b) Patterns of niche overlap among broadly sympatric versus narrowly sympatric Kalahari lizards (Scincidae:*Mabuya*). Ecology 58:121-128

Huey RB, Pianka ER (1981) Ecological consequences of foraging mode. Ecology 62:991-999

Huey RB, Slatkin M (1976) Cost and benefits of lizard thermoregulation. Q Rev Biol 51:363-384

Huey RB, Pianka ER, Egan ME, Coons LW (1974) Ecological shifts in sympatry: Kalahari fossorial lizards (*Typhlosaurus*). Ecology 55:304-316

Huey RB, Pianka ER, Hoffman JA (1977) Seasonal patterns of thermoregulatory behaviour and body temperature of diurnal Kalahari lizards. Ecology 58:1066-1075

Huey RB, Bennett AF, John-Adler H, Nagy KA (1984) Locomotor capacity and foraging behaviour of Kalahari lacertid lizards. Anim Behav 32:41-50

Humphreys WF (1975) The influence of burrowing and thermoregulatory behaviour on the water relations of *Geolycosa godeffroyi* (Araneae:Lycosidae), an Australian wolf spider. Oecologia (Berl) 21:291-311

Humphreys WF (1978) The thermal biology of *Geolycosa godeffroyi* and other burrow inhabiting Lycosidae (Araneae) in Australia. Oecologia (Berl) 31:319-347

Hutchison VH, Kosh RJ (1974) Thermoregulatory function of the parietal eye in the lizard *Anolis carolinesis*. Oecologia (Berl) 16:173-177

Hutchison VH, Dowling HG, Vinegar A (1966) Thermoregulation in a brooding female Indian python, *Python molurus bivittatus*. Science (NY) 151:694-696

Jackson JF (1974) Utilization of periods of high sensory complexity for site change in two lizards. Copeia 1974(3):785-787

Jones SC, Nutting WL (1989) Foraging ecology of subterranean termites in the Sonoran Desert. pp 79-106 in Schmidt (1989)

Judd FW (1975) Activity and thermal ecology of the keeled earless lizard, *Holbrookia propinqua*. Herpetologica 31:137-150

Kaufmann R, Pough FH (1982) The effect of temperature upon the efficiency assimilation of preformed water by the desert iguana (*Dipsosaurus dorsalis*). Comp Biochem Physiol 72A:221-224

Kaufmann T (1966) Observations on some factors which influence aggregation by *Blaps sulcata* (Coleoptera:Tenebrionidae) in Israel. Ann Entomol Soc Am 59:660-664

Kay FR (1972) Activity patterns of *Callisaurus draconoides* at Saratoga Springs, Death Valley, California. Herpetologica 28:65-69

King D (1980) The thermal biology of free-living sand goannas (*Varanus gouldii*) in southern Australia. Copeia 1980(4):755-767

King D, Green B (1979) Notes on diet and reproduction of the sand goanna *Varanus gouldii resenbergi*. Copeia 1979(1):64-70

King WW, Hadley NF (1979) Water flux and metabolic rates of free-roaming scorpions using the doubly-labeled water technique. Physiol Zoöl 52:176-189

Knülle W, Rudolph D (1983) Humidity relationships and water balance of ticks. In: Obenchain FD, Galun R (eds) Physiology of ticks. Pergamon, Oxford, pp 43-70

Kobayashi D, Mautz WJ, Nagy K (1983) Evaporative water loss: humidity acclimation in *Anolis carolinesis* lizards. Copeia 1983(3):701-704

Koch C (1961) Some aspects of abundant life in the vegetationless sand of the Namib Desert dunes. J S W Afr Sci Soc 15:8-34

Koch LE (1978) A comparative study of the structure, function and adaptation to different habitats of burrows in the scorpion genus *Urodacus* (Scorpionida, Scorpionidae). Rec West Aust Mus 6:119-146

Koch LE (1981) The scorpions of Australia: aspects of their ecology and zoogeography. In: Keast A (ed) Ecological biogeography of Australia. Junk, The Hague, pp 875-884

Kofron DP (1983) Female reproductive cycle of the Neotropical snail-eating snake *Sibon sanniola* in northern Yucatan, Mexico. Copeia 1983(4):963-969

Kosh RJ, Hutchison VH (1972) Thermal tolerances of parietalectomized *Anolis carolinesis* acclimatized at different temperatures and photoperiods. Herpetologica 28:183-191

Krakauer T, Gans C, Paganelli CV (1968) Ecological correlations of water loss in burrowing reptiles. Nature 218:659-660

Kramm RA, Kramm KR (1972) Activities of certain species of *Eleodes* in relation to season, temperature, and time of day at Joshua Tree National Monument (Coleoptera:Tenebrionidae). S W Nat 16:341-355

Krapf D (1986) Contact chemoreception of prey in hunting scorpions (Arachnida:Scorpiones). Zool Anz 217:119-129

Krekorian C O'N (1977) Homing in the desert iguana *Dipsosaurus dorsalis*. Herpetologica 33:123-127

Kühnelt W (1975) Beiträge zur Kenntnis der Nahrungsketten in der Namib (Südwestafrika). Verh Ges Ökologie, Wien, pp 197-210

Lamoral BH (1978) Soil hardness, an important and limiting factor in burrowing scorpions of the genus *Opisthophthalmus* CL Koch 1837 (Scorpionidae, Scorpionida). In: Merrett P (ed) Arachnology. Seventh International Congress. Symp Zool Soc Lond 42:171–181

Lancaster J, Lancaster N, Seely MK (1984) Climate of the central Namib desert. Madoqua 14:5-61

Larmuth J (1978) Temperatures beneath stones used as daytime retreats by desert animals. J Arid Environ 1:35-40

Larmuth J (1979) Aspects of plant habit as a thermal refuge for desert insects. J Arid Environ 2:322-327

Lee RE Jr, Baust JG (1985) Low temperature acclimation in the desert spider, *Agelenopsis aperta* (Araneae, Agelenidae). J Arachnol 13:129-136

Lemire M, Vernet R (1982) Effects of different loads on the nasal gland secretion of the lizard *Uromastix acanthinurus* (Agamidae) and relations with its function. Comp Biochem Physiol 72A:65-72

Lemire M, Vernet R (1983) Acclimation of nasal salt excretion in the lizard *Uromastix acanthinurus* (Agamidae). Effects of repetitive salt loads. J Arid Environ 6:145-153

Lemire M, Vernet R, Grenot C (1980) Electrolyte excretion by the nasal gland of an herbivorous Saharan lizard, *Uromastix acanthinurus* (Agamidae). Effects of single NaCl and KCl loads. J Arid Environ 3:325-330

Leuck BE (1982) Comparative burrow use and activity patterns of parthenogenetic and bisexual whiptail lizards (*Cnemidophorus*:Teiidae). Copeia 1982(2):416-424

Lewis TH (1949) Dark coloration in the reptiles of the Tularosa malpais, New Mexico. Copeia 1949(1):181-184

Lewis TH (1951) Dark coloration in the reptiles of the malpais of the Mexican border. Copeia 1951(2):311-312

Licht P (1968) Responses of the thermal preferendum and heat resistance to thermal acclimation under different photoperiods in the lizard *Anolis carolinensis*. Am Midl Nat 79:149-158

Licht P (1972) Environmental physiology of reptilian breeding cycles: role of temperature. Gen Comp Endocrinol (Suppl)3:477-488

Licht P, Bennett AF (1972) A scaleless snake: tests of the role of reptilian scales in water loss and heat transfer. Copeia 1972(4):702-707

Lillywhite HB, Maderson PFA (1982) Skin structure and premeability. In: Gans C, Pough FH (eds) Biology of the Reptilia 12 (Physiology C). Physiological ecology. Academic Press, London, pp 397-442

Linsenmair KE (1968) Anemomenotaktische Orientierung bei Skorpionen (Chelicerata, Scorpiones). Z Vergl Physiol 60:445-449

Linsenmair KE (1984) Comparative studies on the social behaviour of the desert isopod *Hemilepistus reaumuri* and of a *Porcellio* species. Symp Zool Soc Lond 53:423-453

Losos JB (1988) Thermoregulatory correlates of escape behavior by a desert lizard, *Ctenophorus isolepis*. J Herpetol 22:353-356

Louw GN (1972) The role of advective fog in the water economy of certain Namib Desert animals. In: Maloiy GMO (ed) Comparative physiology of desert animals. Symp Zool Soc Lond No 31:297-314

Louw GN, Hamilton III WJ (1972) Physiological and behavioural ecology of the ultra-psammophilous Namib Desert tenebrionid *Lepidochora argentogrisea*. Madoqua (11)1:87-95

Louw GN, Holm E (1972) Physiological, morphological and behavioural adaptations of the ultrapsammophilous, Namib Desert lizard *Aporosaura anchietae* (Bocage). Madoqua (11)1:67-85

Louw GN, Nicolson SW, Seely MK (1986) Respiration beneath desert sand: carbon dioxide diffusion and respiratory patterns in a tenebrionid beetle. J Exp Biol 120:443-447

Louw G, Seely M (1982) Ecology of desert organisms. Longman, London

Louw S (1983) The diversity and daily and seasonal activity of ground-living Tenebrionidae (Coleoptera) in the southern Namib and Kalahari ecosystems. Cimbebasia (A)7(3):35-56

Lowe CH, Heed WB, Halpern EA (1967) Supercooling of the saguaro species *Drosophila nigrospiracula* in the Sonoran Desert. Ecology 48:984-985

Lowe CH, Lardner PJ, Halpern EA (1971) Supercooling in reptiles and other vertebrates. Comp Biochem Physiol 39A:125-135

Luke C (1986) Convergent evolution of lizard toe fringes. Biol J Linn Soc 27:1-16

Mackay RS (1968) Observations of peristaltic activity versus temperature and circadian rhythms in undisturbed *Varanus fluorescens* and *Ctenosaura pectinata*. Copeia 1968(2):252-259

MacKay WP, Zak JC, Whitford WG (1989) The natural history and role of subterranean termites in the northern Chihuahuan Desert. pp 53-77 in Schmidt (1989)

Main BY (1976) Spiders. Collins, Sydney (The Australian Naturalist Library)

Main BY (1982) Adaptations to arid habitats by mygalomorph spiders. In: Barker WR, Greenslade PJM (eds) Evolution of the flora and fauna of arid Australia. Peacock Publications, Frewville South Australia, pp 273-283

Marcellini DL (1976) Some aspects of the thermal ecology of the gecko *Hemidactylus frenatus*. Herpetologica 32:341-345

Marcuzzi G (1960) Rapporti tra equilibrio idrico e ambiente nei Colotteri Tenebrionidi. Arch Zool (Ital) Napoli 45:325-342

Marino PC (1986) Activity patterns and microhabitat selection in a desert tenebrionid beetle (Coleoptera:Tenebrionidae). Ann Entomol Soc Am 79:468-471

Marsh AC (1985) Thermal responses and temperature tolerance in a diurnal desert ant, *Ocymyrmex barbiger*. Physiol Zool 58:629-636

Marsh AC (1987a) Thermal responses and temperature tolerance of a desert ant-lion larva. J Therm Biol 12:295-300

Marsh AC (1987b) The foraging ecology of two Namib Desert harvester ant species. S Afr Tydskr Dierk 22:130-136

Marsh AC (1988) Activity patterns of some Namib Desert ants. J Arid Environ 14:61-73
Mautz WJ (1980) Factors influencing evaporative water loss in lizards. Comp Biochem Physiol 67A:429-437
Mautz WJ (1982) Patterns of evaporative water loss. In: Gans C, Pough FH (eds) Biology of the Reptilia 12 (Physiology C). Physiological ecology. Academic Press, London, pp 443-481
Mautz WJ, Case TJ (1974) A diurnal activity cycle in the granite night lizard, *Xantusia hanshawi*. Copeia 1974(1):243-251
May ML (1979) Insect thermoregulation. Annu Rev Entomol 24:313-349
May ML (1985) Thermoregulation. In: Kerkut GA, Gilbert LI (eds) Comprehensive insect physiology biochemistry and pharmacology, Vol 4. Pergamon, Oxford, pp 507-552
Mayhew WW (1962) Temperature preferences of *Sceloporus orcutti*. Herpetologica 18:217-233
Mayhew WW (1963) Observations on captive *Amphibolurus pictus* an Australian agamid lizard. Herpetologica 19:81-88
Mayhew WW (1964) Photoperiodic responses in three species of the lizard genus *Uma*. Herpetologica 20:95-113
Mayhew WW (1966a) Reproduction in the psammophilous lizard *Uma scoparia*. Copeia 1966(1):14-22
Mayhew WW (1966b) Reproduction in the arenicolous lizard *Uma notata*. Ecology 47:9-18
Mayhew WW (1967) Comparative reproduction in three species of the genus *Uma*. In: Milstead WW (ed) Lizard ecology: a symposium. University of Missouri Press, Columbia, pp 45-61
Mayhew WW (1968) Biology of desert amphibians and reptiles. In: Brown GW Jr (ed) Desert biology, Vol 1. Academic Press, New York, pp 195-356
Mayhew WW, Weintraub JD (1971) Possible acclimatization in the lizard *Sceloporus orcutti*. J Physiol 63:336-340
Mayhew WE, Wright SJ (1971) Water impermeable skin of the lizard *Phrynosoma m'calli*. Herpetologica 27:8-11
McClain E, Praetorius RL, Hanrahan SA, Seely MK (1984) Dynamics of the wax bloom of a seasonal Namib Desert tenebrionid, *Cauricara phalangium* (Coleoptera:Adesmiini) Oecologia (Berl) 63:314-319
McClain E, Seely MK, Hadley NF, Gray V (1985) Wax blooms in tenebrionid beetles of the Namib desert: correlations with environment. Ecology 66:112-118
McCormick SJ, Polis GA (1990) Prey, predators, and parasites. pp 294-320 in Polis (1990a)
McGinnies WG (1979) Arid-land ecosystems — common features throughout the world. In: Goodall DW, Perry RA, Howes KMW (eds) Arid-land ecosystems: structure, functioning and management, Vol 1. Cambridge University Press, Cambridge, pp 299-316
McGinnies WG, Goldman BJ, Paylore P (eds) (1968) Deserts of the world. An appraisal of research into their physical and biological environments. University of Arizona Press, Tucson Arizona
McGinnis SM (1966) *Sceloporus occidentalis*: preferred body temperature of the western fence lizard. Science (NY) 152:1090-1091
McGinnis SM (1970) Flexibility of thermoregulatory behavior in the western fence lizard *Sceloporus occidentalis*. Herpetologica 26:70-76
McGinnis SM, Falkenstein M (1971) Thermoregulatory behavior in three sympatric species of iguanid lizards. Copeia 1971(3):552-554
McGinnis SM, Voigt WG (1971) Thermoregulation in the desert tortoise *Gopherus agassizii*. Comp Biochem Physiol 40A:119-126
McGovern GM, Mitchell JC, Knisley CB (1984) Field experiments on prey selection by the whiptail lizard, *Cnemidophorus inornatus*, in Arizona. J Herpetol 18:327-349
McKenna TM, Packard GC (1975) Rates of heat exchange in the lizards *Cnemidophorus sexlineatus* and *Sceloporus undulatus*. Copeia 1975(1):162-169
McLachlan A (1983) Life-history tactics of rain-pool dwellers. J Anim Ecol 52:545-561
Medica PA (1967) Food habits, habitat preference, reproduction, and diurnal activity in four sympatric species of whiptail lizards (*Cnemidophorus*) in south central New Mexico. Bull S Calif Acad Sci 66:251-276
Medica PA, Arndt RG (1976) Opportunistic feeding in *Sceloporus horridus* from Jalisco, Mexico. Great Basin Nat 36:108-110

Medica PA, Bury RB, Luckenbach RA (1980) Drinking and construction of water catchments by the desert tortoise *Gopherus agassizii*, in the Mojave desert. Herpetologica 36:301-304

Medica RA, Bury RB, Turner FB (1975) Growth of the desert tortoise *Gopherus agassizii* in Nevada. Copeia 1975(4):639-643

Mellini E (1976a) Attività della entomofauna nelli oasi di Mizda e di el-Ghéria esc-Scerghia in primavera avanzata. Resultati delle missioni entomologiche dei professori G. Fiori ed E Mellini nel Nord Africa. Boll Inst Entomol Univ Bologna 33:55-114

Mellini E (1976b) Etologia degli insetti dello Uàdi Sofeggin ed altri Uidiàn della Ghibla nel mese di maggio. Resultati delle missioni entomologiche dei professori G Fiori ed E Mellini nel Nord-Africa. Boll Inst Entomol Univ Bologna 33:115-214

Mendelssohn H (1963) On the biology of the venomous snakes of Israel, Part 1. Isr J Zool 12:143-170

Messina G, Chelazzi G, Chelazzi L, Ercolini A, Ferrara F, Messeri P, Pardi L, Vannini M (1977) Researches on the coast of Somalia. The shore and the dune of Sar Uanle 12. Physical environment: microclimate and soil. Monit Zool Ital (NS) (Suppl) 9:147-181

Miller CM (1944) Ecological relations and adaptations of the limbless lizards of the genus *Anniella*. Ecol Monogr 14:271-289

Minnich JE (1970a) Water and electrolyte balance of the desert iguana, *Dipsosaurus dorsalis* in its natural habitat. Comp Biochem Physiol 35:921-933

Minnich JE (1970b) Evaporative water loss from the desert iguana *Dipsosaurus dorsalis*. Copeia 1970(3):575-578

Minnich JE (1972) Excretion of urate salts by reptiles. Comp Biochem Physiol 41A:535-549

Minnich JE (1979) Reptiles. In: Maloiy GMO (ed) Comparative physiology of osmoregulation in animals, Vol 1. Academic Press, New York, pp 391-641

Minnich JE (1982) The use of water. In: Gans C, Pough FH (eds) Biology of the Reptilia 12 (Physiology C). Physiological ecology. Academic Press, London, pp 325-395

Moehn LD (1974) The effect of quality of light on agonistic behavior of iguanid and agamid lizards. J Herpetol 8:175-183

Moeur JE, Eriksen CH (1972) Metabolic responses to temperature of a desert spider *Lycosa* (*Pardosa*) *carolinesis* (Lycosidae). Physiol Zoöl 45:290-301

Monro J (1967) The exploitation and conservation of resources by populations of insects. J Anim Ecol 36:531-547

Montanucci RR (1971) Ecological and distributional data on *Crotaphytus reticulatus* (Sauria:Iguanidae). Herpetologica 27:183-197

Montanucci RR (1978) Dorsal pattern polymorphism and adaptation in *Gambelia wislizeni* (Reptilia, Lacertilia, Iguanidae). J Herpetol 12:73-81

Moore RF (1978) Seasonal and daily activity patterns and thermoregulation in the southwestern speckled rattlesnake (*Crotalus mitchelli pyrrhus*) and the Colorado desert sidewinder (*Crotalus cerastes laterorepens*). Copeia 1978(3):439-442

Mosauer W (1932) Adaptive convergence in the sand reptiles of the Sahara and of California. Copeia 1932(2):72-78

Mosauer W (1936) The toleration of solar heat in desert reptiles. Ecology 17:56-66

Mueller CF (1970) Temperature acclimation in two species of *Sceloporus*. Herpetologica 26:83-85

Muma MH (1966) Burrowing habits of North American Solpugida (Arachnida). Psyche 73:251-260

Muma MH (1967) Basic behaviour of North American Solpugida. Fla Entomol 50:115-123

Munsey LD (1972) Water loss in five species of lizards. Comp Biochem Physiol 43A:781-794

Murray GW, Schramm D (1987) A comparative study of the diet of the wedge-snouted sand lizard, *Meroles cuneirostris* (Strauch), and the sand diving lizard, *Aporosaura anchietae* (Bocage) (Lacertidae) in the Namib Desert. Madoqua 15:55-61

Muth A (1977a) Body temperatures and associated postures of the zebra-tailed lizard *Callisaurus draconoides*. Copeia 1977(1):122-125

Muth A (1977b) Thermoregulatory postures and orientation to the sun: a mechanistic evaluation for the zebra-tailed lizard *Callisaurus draconoides*. Copeia 1977(4):710-720

Muth A (1980) Physiological ecology of desert iguana (*Dipsosaurus dorsalis*) eggs: temperature and water relations. Ecology 61:1335-1343

Muth A (1981) Water relations of desert iguana (*Dipsosaurus dorsalis*) eggs: temperature and water relations. Physiol Zool 54:441-451

Nagy KA (1972) Water and electrolyte budgets of a free-living desert lizard *Sauromalus obesus*. J Comp Physiol 79:39-62

Nagy KA (1973) Behavior, diet and reproduction in a desert lizard *Sauromalus obesus*. Copeia 1973(1):93-102

Nagy KA (1982) Field studies of water relations. In: Gans C, Pough FH (eds) Biology of the Reptilia 12 (Physiology C). Physiological ecology. Academic Press, London, pp 483-501

Nagy KA (1988) Seasonal patterns of water and energy balance in desert vertebrates. J Arid Environ 14:201-210

Nagy KA, Degen AA (1988) Do desert geckos conserve energy and water by being nocturnal? Physiol Zool 61:495-499

Nagy KA, Knight MH (1989) Comparative field energetics of a Kalahari skink (*Mabuya striata*) and gecko (*Pachydactylus bibroni*). Copeia 1989(1):13-17

Nagy KA, Medica PA (1977) Seasonal water and energy relations of free-living desert tortoises in Nevada: a preliminary report. In: Trotter M (ed) Desert Tortoise Council Symp, Desert Tortoise Council, San Diego, California, pp 152-157

Nagy KA, Huey RB, Bennett AF (1984) Field energetics and foraging mode of Kalahari lacertid lizards. Ecology 65:588-596

Naidu DG, Hattingh J (1986) Water balance and osmoregulation in *Stips stali*, a nocturnal tenebrionid beetle from the Namib Desert. J Insect Physiol 32:891-896

Nicolson SW (1980) Water balance and osmoregulation in *Onymacris plana*, a tenebrionid beetle from the Namib Desert. J Insect Physiol 26:315-320

Nicolson SW, Hanrahan SA (1986) Diuresis in a desert beetle? Hormonal control of the Malpighian tubules of *Onymacris plana* (Coleoptera:Tenebrionidae). J Comp Physiol B156:407-413

Nicolson SW, Bartholomew GA, Seely MK (1984) Ecological correlates of locomotion speed, morphometrics and body temperature in three Namib Desert tenebrionid beetles. S Afr Tydsk Dierk 19:131-134

Nobel PS, Geller GN (1987) Temperature modelling of wet and dry desert soils. J Ecol 75:247-258

Norris KS (1953) The ecology of the desert iguana *Dipsosaurus dorsalis*. Ecology 34:265-287

Norris KS (1967) Color adaptation in desert reptiles and its thermal relationships. In: Milstead WW (ed) Lizard ecology: a symposium. Univ Missouri Press, Columbia, pp 162-229

Norris KS, Dawson WR (1964) Observations on the water economy and electrolyte excretion of chuckwallas (Lacertilia, *Sauromalus*). Copeia 1964(4):638-646

Noy-Meir I (1973) Desert ecosystems: environment and producers. Annu Rev Ecol Syst 4:25-51

Noy-Meir I (1974) Desert ecosystems: higher trophic levels. Annu Rev Ecol Syst 5:195-213

O'Donnell MJ (1977) Site of water vapour absorption in the desert cockroach, *Arenivaga investigata*. Proc Natl Acad Sci USA 74:1757-1760

O'Donnell MJ (1981) Fluid movements during water vapour absorption by the desert burrowing cockroach *Arenivaga investigata*. J Insect Physiol 27:877-887

O'Donnell MJ (1982) Water vapour absorption by the desert burrowing cockroach, *Arenivaga investigata*: evidence against a solute dependent mechanism. J Exp Biol 96:251-262

Orr Y, Shachak M, Steinberger Y (1979) Ecology of the small spotted lizard (*Eremias guttulata guttulata*) in the Negev Desert (Israel). J Arid Environ 2:151-161

Packard GC, Tracy CR, Roth JJ (1977) The physiological ecology of reptilian eggs and embryos, and the evolution of viviparity within the class Reptilia. Biol Rev 52:71-105

Packard MJ, Packard GC, Boardman TJ (1981) Water balance of parchment shelled eggs of a desert lizard (*Callisaurus draconoides*). Can J Zool 58:2051-2058

Packard MH, Packard GC, Boardman TJ (1982) Structure of eggshells and water relations of reptilian eggs. Herpetologica 38:136-155

Parker WS (1971) Ecological observations on the regal horned lizard (*Phrynosoma solare*) in Arizona. Herpetologica 27:333-338

Parker WS (1972) Aspects of the ecology of a Sonoran desert population of the western banded gecko, *Coleonyx variegatus* (Sauria, Eublepharinae). Am Midl Nat 88:209-224

Parker WS, Pianka ER (1973) Notes on the ecology of the iguanid lizard *Sceloporus magister*. Herpetologica 29:143-152

Parker WS, Pianka ER (1974) Further ecological observations on the western banded gecko, *Coleonyx variegatus*. Copeia 1974(3):521-531

Parker WS, Pianka ER (1975) Comparative ecology of populations of the lizard *Uta stansburiana*. Copeia 1975(4):615-632

Parmenter RR, Parmenter CA, Cheney CD (1989) Factors influencing microhabitat partitioning in arid-land darkling beetles (Tenebrionidae): temperature and water conservation. J Arid Environ 17:57-67

Patchell FC, Shine R (1986) Hinged teeth for hard-bodied prey: a case of convergent evolution between snakes and legless lizards. J Zool (Lond) 208:269-275

Patterson R (1971a) The role of urination in egg predator defense in the desert tortoise (*Gopherus agassizi*). Herpetologica 27:197-199

Patterson R (1971b) Visual cliff perception in tortoises. Herpetologica 27:339-341

Pearson OP (1977) The effect of substrate and skin colour on thermoregulation of a lizard. Comp Biochem Physiol 58A:353-358

Peel RF (1974) Insolation weathering: some measurements of diurnal temperature changes in exposed rocks in the Tibesti region, central Sahara. Z Geomorph (Suppl) 21:19-28

Péfaur JE (1981) Composition and phenology of epigeic animal communities in the Lomas of southern Peru. J Arid Environ 4:31-42

Péfaur JE, López-Tejeda E (1983) Ecological notes on the lizard *Tropidurus peruvianus* in southern Peru. J Arid Environ 6:155-160

Perrin MR, Campbell BS (1980) Some aspects of thermoregulation in three species of southern African tortoise. S Afr J Zool 16:35-43

Pianka ER (1967) On lizard species diversity: North American flatland deserts. Ecology 48:333-351

Pianka ET (1969a) Habitat specificity, speciation and species diversity in Australian desert lizards. Ecology 50:498-502

Pianka ER (1969b) Sympatry of desert lizards (*Ctenotus*) in Western Australia. Ecology 50:1012-1030

Pianka ER (1970a) Notes on the biology of *Varanus gouldii flavirufus*. West Aust Nat 11:141-144

Pianka ER (1970b) Comparative autecology of the lizard *Cnemidophorus tigris* in different parts of its geographic range. Ecology 51:703-720

Pianka ER (1971a) Comparative ecology of two lizards. Copeia 1971(1):129-138

Pianka ER (1971b) Ecology of the agamid lizard *Amphibolurus isolepis* in Western Australia. Copeia 1971(3):527-536

Pianka ER (1972) Zoogeography and speciation of Australian desert lizards: an ecological perspective. Copeia 1972(1):127-145

Pianka ER (1973) The structure of lizard communities. Annu Rev Ecol Syst 4:53-74

Pianka ER (1977) Reptilian species diversity. In: Gans C, Tinkle DW (eds) Biology of the Reptilia 7 (Ecology and behaviour A). Academic Press, London, pp 1-34

Pianka ER (1978) Evolutionary ecology (2nd edn). Harper, Row, New York

Pianka ER (1980) Guild structure in desert lizards. Oikos 35:194-201

Pianka ER (1981) Diversity and adaptive radiations of Australian desert lizards. In: Keast A (ed) Ecological biogeography of Australia. W Junk, The Hague, pp 1377-1392

Pianka ER (1982) Observations on the ecology of *Varanus* in the Great Victorian Desert. West Aust Nat 15:37-44

Pianka ER (1985) Some intercontinental comparisons of desert lizards. Natl Geogr Res 1:490-504

Pianka ER (1986) Ecology and natural history of desert lizards. Analysis of the ecological niche and community structure. Princeton Unversity Press, Princeton

Pianka ER, Giles WF (1982) Notes on the biology of two species of nocturnal skinks, *Egernia inornata* and *Egernia striata*, in the Great Victorian Desert. West Aust Nat 15:44-49

Pianka ER, Huey RB (1978) Comparative ecology, resource utilization and niche segregation among gekkonid lizards in the southern Kalahari. Copeia 1978(4):691-701

Pianka ER, Parker WS (1972) Ecology of the iguanid lizard *Callisaurus draconoides*. Copeia 1972(3):493-508

Pianka ER, Parker WS (1975) Ecology of horned lizards: a review with special reference to *Phrynosoma platyrhinos*. Copeia 1975(1):141-162

Pianka ER, Pianka HD (1970) The ecology of *Moloch horridus* (Lacertilia:Agamidae) in Western Australia. Copeia 1970(1):90-103

Pianka ER, Pianka HD (1976) Comparative ecology of twelve species of nocturnal lizards (Gekkonidae) in the Western Australian desert. Copeia 1976(1):125-142

Pierre F (1958) Écologie et peuplement entomologique des sables vifs du Sahara nord-occidental. Centre National de la Recherche scientifique, Paris

Pietruszka RD (1987) Maxithermy and the thermal biology of an herbivorous dune lizard. J Arid Environ 14:175-185

Polis GA (1979) Prey and feeding phenology of the desert sand scorpion *Paruroctonus mesaensis* (Scorpionidae:Vaejovidae). J Zool (Lond) 188:333-346

Polis GA (1980a) Seasonal patterns and age-specific variation in the surface activity of a population of desert scorpions in relation to environmental factors. J Anim Ecol 49:1-18

Polis GA (1980b) The effect of cannibalism on the demography and activity of a natural population of desert scorpions. Behav Ecol Sociobiol 7:25-35

Polis GA (1981) The evolution and dynamics of intraspecific predation. Annu Rev Ecol Syst 12:225-251

Polis GA (1984) Age structure component of niche width and intraspecific resource partitioning: can age groups function as ecological species? Am Nat 123:541-564

Polis GA (1988) Foraging and evolutionary responses of desert scorpions to harsh environmental periods of food stress. J Arid Environ 14:123-134

Polis GA (ed) (1990a) The biology of scorpions. Stanford University Press, Stanford, California

Polis GA (1990b) Ecology. pp 247-293 in Polis (1990a)

Polis GA, Farley RD (1979a) Behavior and ecology of mating in the cannibalistic scorpion, *Paruroctonus mesaensis* Stahnke (Scorpionida:Vaejovidae). J Arachnol 7:33-46

Polis GA, Farley RD (1979b) Characteristics and environmental determinants of natality, growth and maturity in a natural population of the desert scorpion *Paruroctonus mesaensis*. J Zool (Lond) 187:517-542

Polis GA, Farley RD (1980) Population biology of a desert scorpion: survivorship, microhabitat, and the evolution of a life history strategy. Ecology 61:620-629

Polis GA, Sissom WD (1990) Life history. pp 161-223 in Polis (1990a)

Polis GA, Sissom WD, McCormick SJ (1981) Predators of scorpions: field data and a review. J Arid Environ 4:309-327

Polis GA, Myers C, Quinless M (1986) Burrowing biology and spatial distribution of desert scorpions. J Arid Environ 10:137-146

Porter WP, Gates DM (1969) Thermodynamic equilibria of animals with environment. Ecol Monogr 39:245-270

Porter WP, Norris KS (1969) Lizard reflectivity change and its effect on light transmission through body wall. Science (NY) 163:482-484

Porter WP, Mitchell JW, Beckman WA, De Witt CB (1973) Behavioral implications of mechanistic ecology. Thermal and behavioral modeling of desert ectotherms and their microenvironment. Oecologia (Berl) 13:1-54

Pough H (1966) Ecological relationships of rattlesnakes in southeastern Arizona with notes of other species. Copeia 1966(4):676-683

Pough RH (1969) The morphology of understand respiration in reptiles. Herpetologia 25:216-223

Powell H, Heatwole H, Heatwole M (1978) Winter aggregation of *Leiolopisma guichenoti*. Brit J Herpetol 5:789-791

Punzo F (1982) Tail autotomy and running speed in the lizard *Cophosaurus texanus* and *Uma notata*. J Herpetol 16:329-331

Putnam RW, Murphy RW (1982) Low metabolic rate in a nocturnal desert lizard, *Anarbylus switaki* Murphy (Sauria:Gekkonidae). Comp Biochem Physiol 71A:119-123

Rand AS (1964) Inverse relationship between temperature and shyness in the lizard *Anolis lineatopus*. Ecology 45:863-864

Regal PJ (1966) Thermophilic response following feeding in certain reptiles. Copeia 1966(3):588-590

Regal PJ (1975) The evolutionary origin of feathers. Q Rev Biol 50:35-66

Reichman OJ, Prakash I, Roig V (1979) Food selection and consumption. In: Goodall DW, Perry RA, Howes KMW (eds) Arid-land ecosystems: structure, functioning and management, vol 1. Cambridge University Press, Cambridge, pp 681-716

Remmert H (1981) Body size of terrestrial arthropods and biomass of their populations in relation to the abiotic parameters of their milieu. Oecologia (Berl) 50:12-13

Rice GE, Bradshaw SD (1980) Changes in dermal reflectance and vascularity and their effects on thermoregulation in *Amphibolurus nuchalis* (Reptilia:Agmidae). J Comp Physiol 135:139-146

Ricklefs RE, Cochran D, Pianka ER (1981) A morphological analysis of the structure of communities of lizards in desert habitats. Ecology 62:1474-1483

Riddle WA, Pugach S (1976) Cold hardiness in the scorpion, *Paruroctonus aquilonalis*. Cryobiology 12:248-253

Riddle WA, Crawford CS, Zeitone AM (1976) Patterns of haemolymph osmoregulation in three desert arthropods. J Comp Physiol B112:295-305

Riechert SE (1978) Energy-based territoriality in populations of the desert spider *Agenenopsis aperta* (Gertsch). Symp Zool Soc Lond No 42:211-222

Riedesel ML, Cloudsley-Thompson JA, Cloudsley-Thompson JL (1971) Evaporative thermoregulation in turtles. Physiol Zoöl 44:28-32

Rissing SW, Pollock GB (1989) Behavioral ecology and community organization of desert seed-harvester ants. J Arid Environ 17:167-173

Roberts LA (1968) Water loss in the desert lizard *Uta stansburiana*. Comp Biochem Physiol 27:583-589

Robertson HG, Nicolson SW, Louw GN (1982) Osmoregulation and temperature effects on water loss and oxygen consumption in two species of African scorpion. Comp Biochem Physiol 71A:605-609

Robinson MD, Cunningham AB (1978) Comparative diet of two Namib desert sand lizards. Madoqua 11:41-53

Robinson MD, Hughes DA (1978) Observations on the natural history of Peringuey's adder, *Bitis peringueyi* (Boulanger) (Reptilia:Viperidae). Ann Transvaal Mus 31:189-196

Robson GM, Lambert MRK (1980) Observations on the insect food of some semi-desert lizards in southern Morocco. J Arid Environ 3:141-151

Rose FL, Judd FW (1975) Activity and home range size of the Texas tortoise, *Gopherus berlandieri*, in South Texas. Herpetologica 31:448-456

Roth JJ, Ralph CL (1976) Body temperature of the lizard (*Anolis carolinesis*): effect of parietalectomy. J Exp Zool 198:17-28

Ruby DE (1977) Winter activity in Yarrow's spiny lizard *Sceloporus jarrovi*. Herpetologica 33:322-333

Rutowski RL, Alcock J (1989) Insect mating systems in the Sonoran Desert of North America. J Arid Environ 17:157-165

Rzóska J (1961) Observations on tropical rainpools and general remarks on temporary waters. Hydrobiologia 17:265-286

Rzóska J (1984) Temporary and other waters. In: Cloudsley-Thompson JL (ed) Sahara desert. Pergamon, Oxford (Key Environments), pp 105-114

Safriel U, Ayal Y, Kotler BP, Lubin Y, Olsvig-Whittaker L, Pinshow B (1989) What's special about desert ecology? J Arid Environ 17:125-130

Savitzky AH (1981) Hinged teeth in snakes: an adaptation for swallowing hard-bodied prey. Science (NY) 212:346-349

Savitzky AH (1983) Coadapted character complexes among snakes: fossoriality, piscivory and durophagy. Am Zool 23:379-409

Schall JJ (1977) Thermal ecology of five sympatric species of *Cnemidophorus* (Sauria:Teiidae). Herpetologica 33:261-272

Schall JJ, Pianka ER (1980) Evolution of escape behavior diversity. Am Nat 115:551-566

Schmidt JO (ed) (1989) Special biotic relationships in the arid south west. University of New Mexico Press, Albuquerque (contributions of the Committee on Desert and Arid Zones Research of the American Association for the Advancement of Science 24)

Schmidt PJ, Schmidt JO (1989) Harvester ants and horned lizards predator — prey interactions. pp 25-52 in Schmidt (1989)

Schmidt PJ, Sherbrooke WC, Schmidt JO (1989) The detoxification of ant (*Pogonomyrmex*) venom by a blood factor in horned lizards (*Phrynosoma*). Copeia 1989(3):603-607

Schmidt-Nielsen K (1964) Desert animals. Oxford Univ Press, Oxford
Schmidt-Nielsen K, Bentley PJ (1966) Desert tortoise *Gopherus agassizii*: cutaneous water loss. Science (NY) 154:911
Schmidt-Nielsen K, Dawson WR (1964) Terrestrial animals in dry heat: desert reptiles. In: Dill DB (ed) Handbook of physiology, Sect 4. Am Physiol Soc, Washington DC, pp 467-480
Schoener TW (1971) Theory of feeding strategies. Annu Rev Ecol Syst 2:369-404
Scholtz CH (1989) Unique foraging behaviour in *Pachysoma* (= *Scarabaeus*) *striatum* Castelnau (Coleoptera:Scarabeidae): an adaptation to arid conditions? J Arid Environ 16:305-313
Scholtz CH, Caveney S (1988) Adaptations in trogid carrion beetles to extremely arid conditions. J Arid Environ 15:179-191
Schultz TD, Hadley NF (1987) Structural colors of tiger beetles and their role in heat transfer through the integument. Physiol Zool 60:737-745
Seely MK (1978) The Namib dune desert: an unusual ecosystem. J Arid Environ 1:117-128
Seely MK (1979) Irregular fog as a water source for desert dune beetles. Oecologia (Berl) 42:213-227
Seely M (1985) Predation and environment as selective forces in the Namib Desert. In: Vrba ES (ed) Species and speciation. Transvaal Mus Monogr 4:161-165
Seely MK, Griffin M (1986) Animals of the Namib Desert: interactions with their physical environment. Rev Zool Afr 100:47-61
Seely MK, Hamilton WJ III (1976) Fog catchment sand trenches constructed by tenebrionid beetles, *Lepidochora*, from the Namib Desert. Science (NY) 193:484-486
Seely MK, Mitchell D (1987) Is the substrate environment of the Namib Desert dunes a thermal haven for chthonic beetles? S Afr J Zool 22:57-61
Seely MK, Lewis CJ, O'Brien KA, Shuttle AE (1983) Fog response of tenebrionid beetles in the Namib Desert. J Arid Environ 6:135-143
Seely MK, Mitchell D, Roberts CS, McClain E (1988a) Microclimate and activity of the lizard *Angolosaurus skoogi* on a dune slipface. S Afr Tydskr Dierk 23(2):92-102
Seely MK, Roberts CS, Mitchell D (1988b) High body temperatures of Namib dune tenebrionids — why? J Arid Environ 14:135-143
Seibt V, Wickler W (1988) Why do 'family spiders' *Stegodyphus* (Eresidae), live in colonies. J Arachnol 16:193-198
Seymour RS, Vinegar A (1973) Thermal relations, water loss and oxygen consumption of a North American tarantula. Comp Biochem Physiol 44A:83-96
Shachak M (1980) Energy allocation and life history strategy of the desert isopod *H. reaumuri*. Oecologia (Berl) 45:404-413
Shachak M, Brand S (1983) The relationship between sit and wait foraging strategy and dispersal in the desert scorpion, *Scorpio maurus palmatus*. Oecologia (Berl) 60:371-377
Shachak M, Newton PG (1985) The relationship between brood care and environmental unpredictability in the desert isopod *Hemilepistus reaumuri*. J Arid Environ 9:199-209
Sherbrooke WC, Montanucci RR (1988) Stone mimicry in the round-tailed horned lizard, *Phrynosoma modestum* (Sauria:Iguanidae). J Arid Environ 14:275-284
Shine R (1980) Ecology of the Australian death adder *Acanthophis antarcticus* (Elapidae): evidence for convergence with the Viperidae. Herpetologica 36:281-289
Shine RA (1988) Parental care in reptiles. In: Gans C, Huey RB (eds) Biology of the Reptilia, vol 16 (Ecology B). Defense and life history. Alan R Liss, New York, pp 275-329
Shine R, Schwaner TD (1985) Prey constriction by venomous snakes: a review, and new data on Australian species. Copeia 1985(4):1067-1071
Shoemaker VH (1975) Adaptations to aridity in amphibians and reptiles. In: Vernberg FJ (ed) Physiological adaptation to the environment. Intext Educational Publications, New York, pp 143-154
Shoemaker VH, Nagy KA (1977) Osmoregulation in amphibians and reptiles. Annu Rev Physiol 39:449-471
Shook RS (1978) Ecology of the wolf spider *Lycosa carolinesis* Walckenaer (Araneae:Lycosidae) in a desert community. J Arachnol 6:53-64
Shorthouse DJ, Marples TG (1980) Observations of the burrow and associated behaviour of the arid-zone scorpion *Urodacus yaschenkoi* (Birula). Aust J Zool 28:581-590

Slip DJ, Shine R (1988) Reptilian endothermy: a field study of thermoregulation by brooding diamond pythons. J Zool (Lond) 216:367-378

Slobodchikoff CN, Wismann K (1981) A function of the subelytral chamber of tenebrionid beetles. J Exp Biol 90:109-114

Smith GT (1966) Observations on the life history of the scorpion *Urodacus abruptus* Pocock (Scorpionidae), and an analysis of its home sites. Aust J Zool 14:383-398

Smits AW (1985) Behavioral and dietary responses to aridity in the chuckwalla *Sauromalus hispidus*. J Herpetol 19:441-449

Somma LA, Fawcett JD (1989) Brooding behaviour of the prairie skink *Eumeces septentrionalis*, and its relationship to the hydric environment of the nest. Zool J Linn Soc 95:245-256

Stamps JA (1977a) Social behaviour and spacing patterns in lizards. In: Gans C, Tinkle DW (eds) Biology of the Reptilia 7 (Ecology and behaviour A). Academic Press, London, pp 265-334

Stamps JA (1977b) The function of the survey posture in *Anolis* lizards. Copeia 1977(4):756-758

Stebbins RC (1943) Adaptations of the nasal passages for sand burrowing in the saurian genus *Uma*. Am Nat 77:38-52

Stebbins RC (1948) Nasal structures in lizards with reference to olfaction and conditioning of inspired air. Am J Anat 83:183-222

Stebbins RC (1954) Amphibians and reptiles of western North America. McGraw-Hill, New York

Sturbaum BA (1982) Temperature regulation in turtles. Comp Biochem Physiol 72A:615-620

Sturbaum BA, Riedesel ML (1974) Temperature regulation responses of ornate box turtles, *Terrapene ornata*, to heat. Comp Biochem Physiol 48(1):527-528

Sutton SL, Holdich DH (eds) (1984) The biology of terrestrial isopods. Symp Zool Soc Lond No 53

Templeton JR (1960) Respiration and water loss at the higher temperatures in the desert iguana *Dipsosaurus dorsalis*. Physiol Zoöl 33:136-145

Templeton JR (1964a) Cardiovascular response to temperature in the lizard *Sauromalus obesus*. Physiol Zoöl 37:300-306

Templeton JR (1964b) Nasal salt excretion in terrestrial lizards. Comp Biochem Physiol 11:223-229

Templeton JR (1966) Responses of the lizard nasal salt gland to chronic hypersalemia. Comp Biochem Physiol 18:563-572

Templeton JR (1970) Reptiles. In: Whittow GC (ed) Comparative physiology of thermoregulation, Vol 1. Invertebrates and nonmammalian vertebrates. Academic Press, New York, pp 167-221

Templeton JR (1972) Salt and water balance in desert lizards. In: Maloiy GMO (ed) Comparative physiology of desert animals. Symp Zool Soc Lond 31:61-77

Tevis L Jr, Newell IM (1962) Studies on the biology and seasonal cycle of the giant red velvet mite, *Dinothrombium pandorae* (Acari, Trombidiidae). Ecology 43:497-505

Tinkle DW (1967) The life and demography of the side-blotched lizard, *Uta stansburiana*. Misc Publ Mus Zool Univ Mich 132:1-182

Tinkle DW, Ballinger RE (1972) *Sceloporus undulatus*: a study of the intraspecific comparative demography of a lizard. Ecology 53:570-584

Toft CA (1985) Resource partitioning in amphibians and reptiles. Copeia 1985(1):1-21

Tollestrup K (1981) The social behavior and displays of two species of horned lizards, *Phrynosoma platyrhinos* and *Phrynosoma coronatum*. Herpetologica 37:131-141

Toolson EC, Hadley NF (1977) Cuticular permeability and epicuticular lipid composition in two Arizona vejovid scorpions. Physiol Zoöl 50:323-330

Toolson EC, Hadley NF (1979) Seasonal effects on cuticular permeability and epicuticular permeability and epicuticular lipid composition in *Centruroides sculpturatus* Ewing 1928 (Scopiones:Buthidae). J Comp Physiol 129:319-325

Tracy CR (1982) Biophysical modeling in reptilian physiology and ecology. In: Gans C, Pough FH (eds) Biology of the Reptilia 12 (Physiology C). Physiological ecology. Academic Press, London, pp 275-321

Tschinkel WR (1975) A comparative study of the chemical defensive system of tenebrionid beetles: chemistry of the secretions. J Insect Physiol 21:753-783

Turner FB, Medica PA (1982) The distribution and abundance of the flat-tailed horned lizard (*Phrynosoma mcallii*). Copeia 1982(4):815-823

Turner FB, Hoddenback GA, Medica PA, Lannom JR (1970) The demography of the lizard *Uta stansburiana* Baird and Girard in southern Nevada. J Anim Ecol 39:505-519

Turner FB, Medica PA, Kowaleswsky BW (1976) Energy utilization by a desert lizard (*Uta stansburiana*). US/IBP Desert Biome Monogr No 1:57

Turner FB, Medica PA, Bridges KW, Jennrich RI (1982a) A population model of the lizard *Uta stansburiana* in southern Nevada. Ecol Monogr 52:243-259

Turner FB, Medica PA, Jennrich RI, Maza BG (1982b) Frequencies of broken tails among *Uta stansburiana* in southern Nevada and a test of the predation hypothesis. Copeia 1982(4):835-840

Turner FB, Medica PA, Lyons CL (1984a) Reproduction and survival of the desert tortoise (*Scaptochelys agassizii*) in Ivanpah Valley, California. Copeia 1984(4):811-820

Turner FB, Weaver DC, Rorabaugh JC (1984b) Effects of reduction in windblown sand on the abundance of the fringe-toed lizard (*Uma inornata*) in the Coachella valley, California. Copeia 1984(2):370-378

Turner FB, Hayden P, Burge BL, Robertson JB (1986) Egg production by the desert tortoise (*Gopherus agassizii*) in California. Herpetologica 42:93-104

Turner JS, Lombard AT (1990) Body color and body temperature in white and black Namib desert beetles. J Arid Environ 19:303-317

Uvarov B (1977) Grasshoppers and locusts. A handbook of general acridology, Vol 2. Centre for Overseas Pest Research, London

Vannini M, Balzi M, Becciolini A, Carmignani I, Ugolini A (1985) Water exchange between mother and larvae in scorpions. Experientia 41:1620-1621

Vitt LJ (1982) Reproductive tactics of *Ameiva ameiva* (Lacertilia:Teiidae) in a seasonally fluctuating tropical habitat. Can J Zool 60:3133-3120

Vitt LJ (1983a) Ecology of an anuran-eating guild of terrestrial tropical snakes. Herpetologica 39:52-66

Vitt LJ (1983b) Reproduction and sexual dimorphism in the tropical teiid lizard *Cnemidophorus ocellifer*. Copeia 1983(2):359-366

Vitt LJ (1983c) Tail loss in lizards: the significance of foraging and predator escape modes. Herpetologica 39:151-163

Vitt LJ, Congdon JD, Dickson NA (1977) Adaptive strategies and energetics of tail autotomy in lizards. Ecology 58:326-337

Vitt LJ, Goldberg SR (1983) Reproductive ecology of two tropical iguanid lizards: *Tropidurus torquatus* and *Platynotus semitaeniatus*. Copeia 1983(1):131-141

Vitt LJ, Ohmart RD (1974) Reproduction and ecology of a Colorado River population of *Sceloporus magister* (Sauria:Iguanidae). Herpetologica 30:410-17

Vitt LJ, Ohmart RD (1977a) Ecology and reproduction of lower Colorado river lizards: 1. *Callisaurus draconoides* (Iguanidae). Herpetologica 33:214-222

Vitt LH, Ohmart RD (1977b) Ecology and reproduction of lower Colorado river II *Cnemidophorus tigris* (Teiidae) with comparisons. Herpetologica 33:223-234

Vitt LH, Price HJ (1982) Ecological and evolutionary determinants of relative clutch mass in lizards. Herpetologica 38:237-255

Vitt LJ, Van Loben Sels RC, Ohmart RD (1978) Lizard reproduction: annual variation and environmental correlates in the iguanid lizard *Urosaurus graciosus*. Herpetologica 34:241-253

Wagner FH, Graetz RD (1981) Animal-animal interactions. In: Goodall DW, Perry RA, Howes KMW (eds) Arid-land ecosystems: structure, functioning and management, Vol 2. Cambridge University Press, Cambridge, pp 51-83

Wallwork JA (1982) Desert soil fauna. Praeger, New York

Wallwork JA, Kamill BW, Whitford WR (1985) Distribution and diversity patterns of soil mites and other microarthropods in a Chihuahuan desert site. J Arid Environ 9:215-231

Walter H, Breckle SW (1986) Ecological systems of the geobiosphere. 2 Tropical and subtropical zonobiomes (Transl S Gruber). Springer, Berlin Heidelberg New York Tokyo, pp 262-273

Warburg MR (1965a) The influence of ambient temperature and humidity on the body temperature and water loss from two Australian lizards: *Tiliqua rugosa* (Gray) (Scincidae), and *Amphibolurus barbatus* Cuvier (Agamidae). Aust J Zool 13:331-350

Warburg MR (1965b) The evolutionary significance of the ecological niche. Oikos 16:205-213

Warburg MR (1978) Diversity of the herpetofauna in the Mediterranean region of northern Israel. J Arid Environ 1:253-259

Warburg MR, Ben Horin A (1979) Thermal effect on the diel activity rhythm of scorpions from mesic and xeric habitats. J Arid Environ 2:339-346

Warburg MR, Polis GA (1990) Behavioral responses, rhythms and activity patterns, pp 224-246 in Polis (1990a)
Watson RT (1989) Niche separation in Namib Desert dune Lepismatidae (Thysanura:Insecta): detritivores in an allochthonous detritus ecosystem. J Arid Environ 17:37-48
Watson RT, Irish J (1988) An introduction to the Lepismatidae (Thysanura:Insecta) of the Namib Desert sand dunes. Madoqua 15:285-293
Weathers WW (1970) Physiological thermoregulation in the lizard *Dipsosaurus dorsalis*. Copeia 1970(3):549-557
Weathers WW (1971) Some cardiovascular aspects of temperature regulation in the lizard *Dipsosaurus dorsalis*. Comp Biochem Physiol 40(A):503-515
Werner YL (1968) Regeneration frequencies in geckos of two ecological types (Reptilia:Gekkonidae). Vie Milieu 19(1) C:199-221
Werner YL (1969) Eye size in geckos of various ecological types (Reptilia:Gekkonidae and Sphaerodactylidae). Isr J Zool 18:291-316
Werner YL (1972) Temperature effects of inner-ear sensitivity in six species of iguanid lizards. J Herpetol 6:147-177
Werner YL (1983) Temperature effects on cochlear function in reptiles: a personal review incorporating new data. In: Fay RR, Gourevitch G (eds) Hearing and other senses: presentations in honor of EG Wever. Amphora, Groton, pp 149-174
Werner YL (1985) Optimal body temperature for inner-ear performance agrees with field body temperature in *Phelsuma* (Reptilia:Gekkoninae). Herpetol J 1:36-37
Werner YL, Broza M (1969) Hypothetical function of elevated locomotory postures in geckos (Reptilia:Gekkonidae). Isr J Zool 18:349-355
Wharton RA (1982) Observations on the behaviour, phenology and habitat preferences of mydas flies in the central Namib Desert (Diptera:Mydidae). Ann Transvaal Mus 33:145-151
Wharton RA (1987) Biology of the diurnal *Metasolpuga picta* (Kraepelin) (Solifugae, Solpugidae) compared with that of nocturnal species. J Arachnol 14:363-383
Wharton RA, Seely MK (1982) Species composition of and biological notes on Tenebrionidae of the lower Kuiseb River and adjacent gravel plain. Madoqua 13:5-25
Whitfield DC, Livezey RL (1973) Thermoregulatory patterns in lizards. Physiol Zoöl 46:258-296
Whitford WG (ed) (1986) Pattern and process in desert ecosystems. University of New Mexico Press, Albuquerque (Contributions of the Committee on Desert and Arid Zones Research of the American Association for the Advancement of Science:21)
Whitford WG, Bryant M (1979) Behavior of a predator and its prey: the horned lizard (*Phrynosoma cornutum*) and harvester ants (*Pogonomyrmex* spp.). Ecology 60:686-694
Whitford WG, Creusere FM (1977) Seasonal and yearly fluctuations in Chihuahuan desert lizard communities. Herpetologica 33:54-65
Whitford WG, Ettershank G (1975) Factors affecting foraging activity in Chihuahuan desert harvester ants. Environ Entomol 4:689-696
Whitford WG, Freckman DW, Elkins NZ, Parker LW, Parmalee R, Phillips J, Tucker S (1981) Diurnal migration and responses to simulated rainfall in desert soil microarthropods and nematodes. Soil Biol Biochem 13:417-425
Whitman DW (1987) Thermoregulation and daily activity patterns in a black desert grasshopper, *Taeniopoda eques*. Anim Behav 35:1814-1826
Whitman DW (1988) Function and evolution of thermoregulation in the desert grasshopper *Taeniopoda eques*. J Anim Ecol 57:369-383
Whitman DW, Orsak L (1985) Biology of *Taeniopoda eques* (Orthoptera:Acrididae) in southeastern Arizona. Ann Entomol Soc Am 78:811-825
Whittow GG (1973) Evolution of thermoregulation. In: Whittow GC (ed) Comparative physiology of thermoregulation. Vol 3: Special aspects of thermoregulation. Academic Press, New York, pp 201-258
Wickler W (1968) Mimicry in plants and animals. Weidenfeld & Nicolson, London. McGraw-Hill, New York
Wiens JA (1976) Population responses to patchy environments. Annu Rev Ecol Syst 7:81-120
Williams CB (1954) Some bioclimatic observations in the Egyptian desert. In: Cloudsley-Thompson JL (ed) Biology of deserts. Institute of Biology, London, pp 18-24

Williams SC (1966) Burrowing activities of the scorpion *Anuroctonus phaiodactylus* (Wood) (Scorpionida:Vejovidae). Proc Calif Acad Sci (4)34:419-428

Williams SC (1987) Scorpion bionomics. Annu Rev Entomol 32:275-295

Willmer PG (1982) Microclimate and the environmental physiology of insects. Adv Insect Physiol 16:1-87

Withers PC (1981) Physiological correlates of limblessness and fossorality in scincid lizards. Copeia 1981(1):197-204

Witten GJ, Heatwole H (1978) Preferred temperature of the agamid lizard *Amphibolurus nobbi nobbi*. Copeia 1978(2):362-364

Wood TG, Lee KE (1971) Abundance of mounds and competition among colonies of some Australian termite species. Pedobiologia 11:341-366

Wooten RC Jr, Crawford CS, Riddle WA (1975) Behavioural thermoregulation of *Orthoporus ornatus* (Diplopoda:Spirostreptidae) in three desert habitats. Zool J Linn Soc 57:59-74

Worthington RD (1982) Dry and wet year comparisons of clutch and adult body sizes of *Uta stansburiana stejnegeri*. J Herpetol 16:332-334

Zeh DW, Zeh JA (1989) Ovipositors, amnions and eggshell architecture in the diversification of terrestrial arthropods. Q Rev Biol 64:147-168

Subject Index

Page numbers in *italics* refer to figures

MIX
Papier aus verantwortungsvollen Quellen
Paper from responsible sources
FSC® C105338

If you have any concerns about our products,
you can contact us on
ProductSafety@springernature.com

In case Publisher is established outside the EU,
the EU authorized representative is:
Springer Nature Customer Service Center GmbH
Europaplatz 3, 69115 Heidelberg, Germany

Printed by Libri Plureos GmbH
in Hamburg, Germany